Application of HACCP for Distribution System Protection

About the Awwa Research Foundation

The Awwa Research Foundation (AwwaRF) is a member-supported, international, nonprofit organization that sponsors research to enable water utilities, public health agencies, and other professionals to provide safe and affordable drinking water to consumers.

The Foundation's mission is to advance the science of water to improve the quality of life. To achieve this mission, the Foundation sponsors studies on all aspects of drinking water, including supply and resources, treatment, monitoring and analysis, distribution, management, and health effects. Funding for research is provided primarily by subscription payments from approximately 1,000 utilities, consulting firms, and manufacturers in North America and abroad. Additional funding comes from collaborative partnerships with other national and international organizations, allowing for resources to be leveraged, expertise to be shared, and broad-based knowledge to be developed and disseminated. Government funding serves as a third source of research dollars.

From its headquarters in Denver, Colorado, the Foundation's staff directs and supports the efforts of more than 800 volunteers who serve on the board of trustees and various committees. These volunteers represent many facets of the water industry, and contribute their expertise to select and monitor research studies that benefit the entire drinking water community.

The results of research are disseminated through a number of channels, including reports, the Web site, conferences, and periodicals.

For subscribers, the Foundation serves as a cooperative program in which water suppliers unite to pool their resources. By applying Foundation research findings, these water suppliers can save substantial costs and stay on the leading edge of drinking water science and technology. Since its inception, AwwaRF has supplied the water community with more than $300 million in applied research.

More information about the Foundation and how to become a subscriber is available on the Web at **www.awwarf.org.**

Application of HACCP for Distribution System Protection

Prepared by:
Kathy Martel, **Gregory Kirmeyer**, and **Amie Hanson**
HDR/EES, Inc., Bellevue, WA

Melita Stevens
Melbourne Water Corporation, Melbourne, VIC 3001, Australia

Joanne Mullenger
South East Water Ltd., Altona Meadows, VIC, Australia

and

Daniel Deere
Water Futures, LLC, Dundas Valley, NSW, Australia

Jointly sponsored by:
Awwa Research Foundation
6666 West Quincy Avenue, Denver, CO 80235-3098

and

U.S. Environmental Protection Agency
Washington D.C.

Published by:

Awwa Research Foundation

American Water Works Association

IWA Publishing

DISCLAIMER

This study was jointly funded by the Awwa Research Foundation (AwwaRF) and the U.S. Environmental Protection Agency (USEPA) under Cooperative Agreement No. R829409-01. AwwaRF and USEPA assume no responsibility for the content of the research study reported in this publication or for the opinions or statements of fact expressed in the report. The mention of trade names for commercial products does not represent or imply the approval or endorsement of AwwaRF or USEPA. This report is presented solely for informational purposes.

CONTENTS

TABLES

FIGURES

FOREWORD

The Awwa Research Foundation is a nonprofit corporation that is dedicated to the implementation of a research effort to help utilities respond to regulatory requirements and traditional high-priority concerns of the industry. The research agenda is developed through a process of consultation with subscribers and drinking water professionals. Under the umbrella of a Strategic Research Plan, the Research Advisory Council prioritizes the suggested projects based upon current and future needs, applicability, and past work; the recommendations are forwarded to the Board of Trustees for final selection. The foundation also sponsors research projects through the unsolicited proposal process; the Collaborative Research, Research Applications, and Tailored Collaboration programs; and various joint research efforts with organizations such as the U.S. Environmental Protection Agency, the U.S. Bureau of Reclamation, and the Association of California Water Agencies.

This publication is a result of one of these sponsored studies, and it is hoped that its findings will be applied in communities throughout the world. The following report serves not only as a means of communicating the results of the water industry's centralized research program but also as a tool to enlist the further support of the nonmember utilities and individuals.

Projects are managed closely from their inception to the final report by the foundation's staff and large cadre of volunteers who willingly contribute their time and expertise. The foundation serves a planning and management function and awards contracts to other institutions such as water utilities, universities, and engineering films. The funding for this research effort comes primarily from the Subscription Program, through which water utilities subscribe to the research program and make an annual payment proportionate to the volume of water they deliver and consultants and manufacturers subscribe based on their annual billings. The program offers a cost-effective and fair method for funding research in the public interest.

A broad spectrum of water supply issues is addressed by the foundation's research agenda: resources, treatment and operations, distribution and storage, water quality and analysis, toxicology, economics, and management. The ultimate purpose of the coordinated effort is to assist water suppliers to provide the highest possible quality of water economically and reliably. The true benefits are realized when the results are implemented at the utility level. The foundation's trustees are pleased to offer this publication as a contribution toward that end.

Walter J. Bishop
Chair, Board of Trustees
Awwa Research Foundation

Robert C. Renner, P.E.
Executive Director
Awwa Research Foundation

ACKNOWLEDGMENTS

The authors would like to acknowledge the gracious financial support of the Awwa Research Foundation and the U.S. Environmental Protection Agency. Without their support, this project would not have been possible.

This research has been guided from the outset by an experienced Project Advisory Committee (PAC) who provided valuable guidance, expertise and suggestions. The authors wish to thank these individuals for their time and efforts:

Homer Emery, San Antonio Water Systems, Texas
Bruce Macler, EPA Region 9, California
Susan Shaw, U.S. Environmental Protection Agency, Washington D.C.
David Lipsky, NYC DEP/Bureau of Water Supply, New York

Four participating water utilities conducted pilot studies for this research project. Each utility formed a team of experts to develop and implement their HACCP Plan. These utility professionals contributed their time and expertise to this research while performing their other responsibilities. The authors gratefully acknowledge these efforts:

Dan Pedersen, City of Austin Water and Wastewater Utility, Texas
Edward Ojeda, City of Austin Water and Wastewater Utility, Texas
Teresa Lutes, City of Austin Water and Wastewater Utility, Texas
Robert Kuhn, City of Austin Water and Wastewater Utility, Texas
Jane Burazer, City of Austin Water and Wastewater Utility, Texas
Onnie Bohr, City of Austin Water and Wastewater Utility, Texas
Rosie Barrios, City of Austin Water and Wastewater Utility, Texas
Tony Bennett, Texas Commission on Environmental Quality

Mike Nadeau, South Berwick Water District, Maine
John Leach, South Berwick Water District, Maine
Jerry Leavitt, South Berwick Water District, Maine
Rick deRochemont, South Berwick Water District, Maine

Corinna Doolan, Sydney Water Corporation, Australia
Lindsay Mullard, Sydney Water Corporation, Australia
Colin Storey, General Water Australia
Keith Ross, General Water Australia
M. Govintharajah, Sydney Water Corporation, Australia
David Cooper, Sydney Water Corporation, Australia
Shariff Shockair, Sydney Water Corporation, Australia
Tony Venturino, Sydney Water Corporation, Australia
Carl Deininger, Sydney Water Corporation, Australia
Peter Cresta, Sydney Water Corporation, Australia
Philip Broad, Sydney Water Corporation, Australia
Dammika Vitanage, Sydney Water Corporation, Australia

Simon Copley, Power and Water Corporation, Australia
Norm Cramp, Power and Water Corporation, Australia
Mark Ewin, Power and Water Corporation, Australia
Gabrielle Halcrow, Power and Water Corporation, Australia
Paul Heaton, Power and Water Corporation, Australia
Peter Hopkins, Power and Water Corporation, Australia
Noel McCarthy, Power and Water Corporation, Australia
Leon Miles, Power and Water Corporation, Australia
Kevin O'Brien, Power and Water Corporation, Australia
Declan Page, Power and Water Corporation, Australia

Five HACCP-certified utilities in Australia provided case study information on their experiences in implementing HACCP. The authors gratefully acknowledge these efforts:

Bob Gray, Brisbane Water
George Ruta, City West Water
David Smith, Gold Coast Water
Dr. Greg Ryan, South East Water
Asoka Jayaratne, Yarra Valley Water

The authors would like to acknowledge Annette Davison of Water Futures in Australia for her assistance in conducting HACCP training workshops for Sydney Water Corporation and Power and Water Corporation, and for review of Project work products.

The authors thank Julie Self of HDR/EES, Inc. for her efforts in bringing this work together into its final form.

EXECUTIVE SUMMARY

WHAT DOES THE TERM "HACCP" REPRESENT?

HACCP (pronounced as "hass-up") is an acronym for Hazard Analysis and Critical Control Point.

WHAT IS HACCP AND WHAT ARE THE MAIN STEPS?

HACCP is an internationally recognized process control system that involves identifying and prioritizing hazards and risks to the quality of food or drinking water, and controlling processes to reliably maintain the desired level of quality. The application of HACCP in a systematic manner helps the water utility control water quality risks as close to their sources as possible.

The 12 steps of HACCP are defined by the Codex Alimentarius Commission (Codex 1997), an intergovernmental body established to implement the Joint Food and Agricultural Organization (FAO)/World Health Organization (WHO) Food Standards Program. The information prepared in completing these 12 steps constitutes the utility's HACCP Plan. These steps are summarized below:

Step 1	Assemble HACCP Team
Step 2	Describe Drinking Water
Step 3	Identify Intended Use
Step 4	Construct Flow Diagram
Step 5	Confirm Flow Diagram
Step 6	Conduct a Hazard Analysis
Step 7	Determine the Critical Control Points
Step 8	Establish Critical Limit(s)
Step 9	Establish a System to Monitor Control of the Critical Control Points
Step 10	Establish Corrective Actions
Step 11	Validate/Verify HACCP Plan
Step 12	Establish Documentation and Recordkeeping.

The contemporary application of the 12 HACCP steps requires their implementation with a broader HACCP System that includes Supporting Programs. Ideally, the Supporting Programs should be in place prior to embarking on the 12 HACCP steps. Supporting Programs included within HACCP systems in the water industry include:

- Staff training and certification programs
- Distribution system maintenance programs
- Standard operating procedures
- Emergency response program
- Quality assurance programs
- Data management systems
- Customer relations program
- Calibration of monitoring systems.

WHERE DID HACCP ORIGINATE AND HOW HAS IT BEEN APPLIED?

The HACCP concept and acronym was first conceived in the US in 1959 by the Pillsbury Company to improve food safety for manned space missions by the National Aeronautics and Space Administration (NASA). Since the 1980's, the HACCP system has evolved, been improved and widely adopted by the food and beverage industries world-wide where it forms an important part of their "food safety plans." In the US, certain food processors, such as seafood, red meat, and poultry industries, are required by federal regulations to develop and implement a HACCP plan. The International Bottled Water Association requires its members to develop a HACCP program for each of their facilities. Since the mid-1990's, HACCP has been applied by a number of individual drinking water systems in Australia, Iceland, France, Canada, and Switzerland. The HACCP system has been incorporated into many drinking water regulatory requirements and guidelines around the globe. For example, the third edition of the WHO drinking water guidelines (WHO 2004) outline a framework for drinking water safety based on the multiple barrier approach and several risk management and quality management approaches including HACCP.

WHAT WAS THE MAIN OBJECTIVE OF THIS PROJECT?

The purpose of this project was to evaluate the application of HACCP in the drinking water industry to protect and maintain distribution system water quality.

WHAT EVALUATION CRITERIA WERE USED TO ASSESS THE APPLICATION OF HACCP TO DISTRIBUTION SYSTEMS?

Six categories of evaluation criteria were used to assess HACCP implementation during the Project pilot studies and the case studies of full-scale, longer, term HACCP implementation. These categories include:

- **Customer satisfaction:** Has HACCP implementation resulted in reduced customer inquiries or complaints about water quality received by the utility?
- **Regulatory compliance:** Has HACCP implementation resulted in improved water quality compliance for the utility?
- **Risk management:** Has HACCP implementation resulted in strengthened mechanisms for water quality risk management at the utility?
- **System management:** Has HACCP implementation resulted in improved system management and control at the utility?
- **Human factors:** Has HACCP implementation improved the human relationships and understanding within the utility?
- **Costs and benefits:** Has HACCP implementation resulted in benefits that outweigh the costs to the utility?

WHAT WERE THE MAIN FINDINGS FROM THE PROJECT'S PILOT STUDIES?

All four pilot study utilities took part in training workshops and subsequently took their HACCP plan documents to various levels of completion. Two of the four participating utilities were successful in implementing the full scope of their HACCP Plan over a 12-month study period: the City of Austin, Texas and the Katherine system in the Northern Territories of Australia. The implementation of HACCP to water distribution systems was feasible and practical, but the time and resource requirements were greater than originally anticipated by the utilities when they agreed to participate.

For the South Berwick Water District, being a very small utility with only three full-time employees working during the latter part of the pilot, developing and implementing a full HACCP system was not achievable given the many day-to-day duties and activities of the staff. This revealed a fundamental difficulty in the implementation of HACCP, or for that matter, any systematic management system, within very small water utilities. Changes that could be made to enable implementation to be successful in very small utilities were considered to include:

- Increase the number of employees on staff within small utilities to provide a critical mass of personnel.
- Provide temporary, contracted support for long enough to implement the system, funded by a third party, if required.
- Provide generic HACCP plans and guidance combined with very explicit guidance, support and tools to help utilities implement the systems in practice.
- Provide more support from the state or regional regulatory and support organizations.

For the larger utilities, there were sufficient resources to implement HACCP although there was a requirement to have some specific resources reasonably dedicated and assigned to make implementation happen. Importantly, it wasn't the preparation of the HACCP plan per se that was resource-intensive. Rather, it was the implementation of the HACCP plan to create a functional HACCP system in practice that required resources. Therefore, it can be concluded that implementing HACCP, and for that matter, any management system of similar rigor, is likely to be a significant undertaking for a water utility.

All the participating utilities concluded that their participation in the HACCP process was a valuable one. The development of the HACCP plan was useful in honing in on the most important risks and process controls for water quality management. Therefore, the decision about whether or not to implement HACCP comes down to an assessment of whether or not the benefits are worthwhile given the costs.

WHAT WERE THE MAIN FINDINGS FROM THE PROJECT'S CASE STUDIES?

All five case study utilities completed evaluation criteria and many provided additional comments and information on their experiences in implementing HACCP systems. The most important findings from the case studies are summarized below:

- Most utilities that have gained HACCP certification have done so after some core management systems [e.g. International Organization for Standards (ISO) 9001, ISO 14001] had been developed and implemented. These management systems helped the utility to gain management control of people and processes which made implementing HACCP relatively straightforward.
- In practice, the case study utilities did not operate multiple, separate systems for quality management, occupational safety, water quality and safety, and environmental considerations. Although separately identifiable and auditable, in operation, all of these systems were captured within an integrated management system (IMS). The

principal benefit of an IMS, as identified by water utilities, was the avoidance of duplication, leading to reduced staff time and costs, and improved process integration.

- Water quality improvements did become evident following the implementation of HACCP, but in most cases, those changes did not appear conclusive until a consistent pattern of improvement had emerged after three or more years. Water quality improvements included reduced numbers of customer complaints and water quality incidents, and fewer microbial indicators.
- When a utility implements HACCP for water, it is likely to be doing so as one of a number of water quality improvement projects, making it very difficult, if not impossible, to precisely determine the extent to which HACCP contributed to any effect(s).
- All utilities that had implemented HACCP and attained certification continued to be audited and re-registered each year since all believed that, overall, the benefits of the HACCP system, including the certification discipline, outweighed the costs.
- Auditing, though sometimes uncomfortable for operating staff, is a necessary and useful element of HACCP. It forces periodic reviews and keeps utility staff and management up-to-date on important issues.

HOW CAN THIS REPORT BE USED?

This report may be useful to water utilities that are considering whether or not to apply the HACCP system to their distribution system. These readers are referred to the following chapters:

- Chapter 2 describes background information on HACCP and current applications.
- Chapter 3 provides a tailored HACCP guidance manual providing guidance on how to implement HACCP including resource requirements and the need for supporting utility programs and standard operating procedures.
- Chapter 4 presents findings from short term pilot studies (12 months duration) that evaluate HACCP based on six different categories: customer satisfaction, regulatory compliance, risk management, system management, human factors, and costs and benefits.
- In Chapter 5, case study information from HACCP-certified utilities in Australia is used to evaluate HACCP based on the same criteria as the pilot studies in Chapter 4. These case studies also provide insight into the challenges of implementing HACCP.

Example workshop training materials are provided in Appendix A. Example HACCP Plans for the four participating utilities are provided in Appendix B.

For industry researchers that are interested in furthering the knowledge and applications potential for HACCP in drinking water systems, research needs are outlined in Chapter 6.

WHAT ARE THE MAIN BENEFITS OF IMPLEMENTING HACCP?

The HACCP system complements the current US approach to managing distribution system water quality that is centered on water quality monitoring and compliance with regulatory requirements. It will help utilities focus resources on key risks and to improve emphasis on operations processes and controls.

The five case study utilities all reported some benefits in aspects of water quality management that were at least partly related to HACCP. Improvements considered by utilities to be associated with the implementation of HACCP included the following:

Water quality
- Reduced customer complaints
- Reduced water quality incidents
- Reduced microbial indicator failures

Operations
- Improved treatment system performance
- Improved work processes

Business performance
- Improved capability to demonstrate due diligence
- Reduced costs
- Improved documentation and record keeping
- Improved training and culture of operational good practices.

The most notable improvement reported by utilities following implementation of HACCP was an apparent drop in the frequency of customer complaints. For example, water quality complaints dropped steadily following HACCP implementation for Gold Coast Water and South East Water by a total of 66 percent and 60 percent, respectively, over a few years.

Three of the five case study utilities - South East Water, Yarra Valley Water, and Gold Coast Water - found that HACCP implementation has helped to improve system credibility and demonstration of due diligence or the "prevention of foreseeable harm." South East Water attributed these improvements to improved record-keeping practices. At Yarra Valley Water, employees have an improved awareness of water quality risks and management processes. Gold Coast Water noted that the structure of the HACCP system has helped with system credibility and demonstration of due diligence.

WHAT ARE THE COSTS OF IMPLEMENTING HACCP?

The case studies for HACCP-certified utilities in Australia, discussed in Chapter 5, show that HACCP implementation costs were relatively modest, and in some cases were met entirely with existing resources. It is important to note that these utilities had all implemented other quality management systems prior to implementing HACCP and therefore, their costs may not be comparable to a utility implementing only HACCP. The reported cash costs of implementing HACCP varied between utilities:

- One time cash costs for HACCP training, coaching, technical advice and documentation varied from $0 (Brisbane Water) to $70,000 (Australian Dollars) ($55,000 US Dollars) (Yarra Valley Water). The variability is primarily explained by the difference in the extent to which utilities used external service providers to undertake some of the tasks.

- Recurrent cash costs associated with increased water quality monitoring and instrumentation varied from $0 (Brisbane Water) through to $105,000 (Australian Dollars) ($84,000 US Dollars) (Gold Coast Water) annually. The variability is related to the extent to which additional monitoring or operational costs arose from the increased risk management activities deemed appropriate by the HACCP team.
- Some utilities, such as Yarra Valley Water, adopted the recommended practice of annual HACCP awareness training, requiring an additional recurrent cash cost of $1,500 (Australian Dollars) ($1,200 US Dollars) annually. New staff, or staff that are losing familiarity with HACCP, attend the training course each year.

Non-cash costs of HACCP implementation varied between utilities. Implementation almost always required a full time, dedicated officer, for at least one month and up to twelve months. For example, City West Water required 0.75 a full-time equivalent (FTE) for one month, while Gold Coast Water utilized 1 FTE for twelve months. The variability is related to the extent to which tasks were split between a core coordinator and other staff, as well as the amount of work required to implement the HACCP plan. For example, City West Water had completed the vast majority of tasks required by HACCP before the organization formally began to undertake its HACCP process and was able to complete the HACCP plan within one month.

ARE THERE ANY SPECIAL CONSIDERATIONS FOR SMALL SYSTEMS?

Small systems often lack expertise in one or more topic areas covered by the HACCP Plan. Outside experts may be utilized as necessary to help facilitate the HACCP training workshop and to support the utility HACCP team.

The small utility may lack adequate historical water quality and system data to identify and rank risks/hazards in the HACCP hazard analysis (Step 6). This dilemma, experienced by the South Berwick Water District as part of their Project-pilot study, was addressed by initially focusing the HACCP Plan to collect additional information to evaluate, document, and improve control over these hazards.

In practice, the small utility may lack adequate manpower or other necessary resources to develop and implement a HACCP Plan independently. This constraint can be addressed but in some cases, the solutions are not directly deliverable by the small utility acting alone (see above notes on findings from pilot studies).

ARE THERE ANY SPECIAL CONSIDERATIONS FOR LARGE SYSTEMS?

Large water utilities face different types of constraints than small systems in implementing HACCP. Communication across departments and between management and operational staff can be challenging. Organization-wide coordination on water quality management issues can also be difficult. These challenges can be addressed through the HACCP training workshop and through careful selection of HACCP team members and, most importantly, through strong, proactive support from the highest levels in the organization.

WHAT STRATEGIES ARE USED TO SUCCESSFULLY IMPLEMENT A HACCP PLAN?

Prior to initiating a HACCP Plan, a utility should complete the following tasks:

- Identify goals and expected benefits.
- Estimate and commit resources required.
- Solicit employee support and commitment.
- Conduct HACCP training workshop.
- Identify and update supporting programs and practices.

The identification of goals and expected benefits helps the utility to focus resources most efficiently and to help sell the program to utility management, staff, and customers. The utility should estimate the resource requirements for implementing HACCP in their system and make a firm commitment to providing these resources prior to initiating the process.

To assure the effectiveness of the HACCP system, it is important to gain the support of and commitment from all employees, from the front line operators to the top managers. The utility HACCP team should include experienced personnel across a wide range of topic areas to assure that all hazards can be identified and properly managed. Before HACCP implementation, utility staff should be formally trained in HACCP terminology and methodology.

Prior to initiating a HACCP plan, it is important to identify other existing utility programs and SOPs that may contribute valuable information related to the condition of the distribution system and distribution system water quality. The recommended approach is to improve water quality programs and practices to meet the AWWA Standard G200-Distribution System Operation and Management (AWWA 2004).

HOW OFTEN DOES THE HACCP PLAN NEED TO BE UPDATED?

The HACCP Plan is considered to be a "live" document that is continually updated as system changes occur, such as physical improvements or operational modifications that affect the process flow diagram. Change is inevitable as the utility implements the HACCP Plan to address the initial priority hazards by instituting appropriate control measures, monitoring, and documentation. Examples of physical improvements that should be documented in the HACCP Plan include addition of new disinfection facilities or pump stations in the distribution system, and covering an open reservoir. Examples of operational modifications that should be documented in the HACCP Plan include changes in disinfectant dosage rate, changes to pressure zone boundaries or flow direction, and use of new standard operating procedures once the initial priority hazards are controlled, the utility should review the hazard analysis fully (HACCP Step 6) and modify the HACCP Plan to focus on the next set of priority hazards.

As new drinking water regulations are promulgated or existing regulations are modified, new requirements should be documented in the HACCP Plan (HACCP Step 2).

If the utility is maintaining HACCP registration or certification, the HACCP Plan should be reviewed and updated in preparation for the independent audit which may occur semiannually or annually depending on the specific registration requirements.

WHAT ARE THE RECOMMENDED NEXT STEPS FOR APPLYING HACCP TO WATER UTILTIES?

The US drinking water industry is somewhat lagging behind its counterparts overseas in applying HACCP to watershed management, water treatment facilities and distribution systems. Many European and Australian utilities have implemented HACCP systems to full certification or have risk management or quality assurance programs that are more or less equivalent, albeit differently labeled. Although several HACCP applications in Australia and Ontario, Canada were motivated by waterborne disease outbreaks or suspected contamination incidents, a superior approach would be to develop and implement quality assurance programs, such as HACCP prior to experiencing such an outbreak or contamination event.

A recommended first step is to improve utility programs and practices to meet the AWWA Standard G200-Distribution Systems Operation and Maintenance. The specifications for this standard include many water utility programs and practices that are needed to successfully implement HACCP. In particular, the AWWA Standard-G200 emphasizes the need for documentation and record-keeping for all programs, key elements of the HACCP approach. Once this step is accomplished, the utility may consider developing and implementing the ISO management systems (ISO 9001 and ISO 14001) or an integrated management system that includes HACCP as well as the ISO systems.

Other recommended steps include:

- Consider whether the HACCP water quality risk management framework could be applied beneficially within the US regulatory system.
- Develop further support and case studies for HACCP application to small water utilities.

CHAPTER 1
BACKGROUND INFORMATION

INTRODUCTION

Current distribution system management practices may leave some systems vulnerable to contamination, as evidenced by failures linked to waterborne disease outbreaks. US waterborne disease records from 1920 to the present show that up to 40 percent of waterborne disease outbreaks have been caused by distribution system problems (Lippy and Waltrip 1984; Kramer et al. 1996; Levy et al. 1998). For example, a *Salmonella typhimurium* outbreak in Gideon, Missouri, was likely caused by poor distribution system flushing practices that triggered the complete draining of two finished water storage facilities into the system. The stored water was contaminated by bird droppings that carried the *Salmonella typhimurium* pathogen (Clark et el. 1997; Geldreich 1996). An outbreak of hemorrhagic *Escherichia coli* (E. coli) serotype O157:H7 occurred in Cabool, Missouri during December 1989 and January 1990 and resulted in 243 cases of diarrhea and 4 deaths (Swerdlow et al. 1992). Shortly before the peak of the outbreak, 45 water meters were replaced and two water mains ruptured. Swerdlow et al. (1992) concluded that system-wide chlorination as well as hyperchlorination during repairs might have prevented this outbreak.

Inadequacies in design, operation and maintenance of water distribution systems have also been documented through recent research funded by the American Water Works Association Research Foundation (AwwaRF). Kirmeyer et al. (2001) identified various potential pathways through which contaminants may enter the distribution system, including poorly designed, constructed, or maintained storage facilities, unprotected cross-connections; and water main break and repair sites. Kirmeyer et al. (2001) also documented evidence of fecal pollution and enteric viruses in the vicinity of water main repair sites, confirming the potential for pathogen intrusion given an available pathway and favorable pressure conditions. Further research by Friedman et al. (2004) on pathogen intrusion into distribution systems during pressure transients confirmed that transient negative pressures do occur in distribution systems and that significant volumes of water have been shown to enter a pilot-scale distribution system through small leaks during transient pressure events. Based on surveys of tank inspection firms, State primacy agencies, and utilities, Kirmeyer et al. (1999) concluded that many storage facilities in the US are not ever inspected and many facilities are inspected less frequently than the three- to five-year frequency recommended by the American Water Works Association (AWWA). A comprehensive survey on cross-connection control programs (Lee et al. 2003) showed that 91 percent of survey respondents have developed cross-connection control programs, but only 49 percent have a requirement for reporting backflow incidents (to state primacy agencies). There is no Federal reporting requirement for cross-connections. In Lee's survey, most respondents were community water systems (99 percent of those surveyed).

In addition to inadequacies in water system processes and procedures, the human element may affect a water system's susceptibility to contamination. Kuslikis and White (2004) provide two examples of how water system employees can impact risks and risk management:

- Competent and loyal employees sometimes take shortcuts to save time or money and unknowingly take major risks.

- Employees may not see the big picture and may have a much higher tolerance of risk than the corporation.

Smith (2004a) also acknowledges that some utility personnel may occasionally be careless, poorly trained, or may lack adequate resources to do their job. Most often, when things go wrong, nothing happens. But to be assured that the system is well-protected from risks, Smith (2004a) advises that a utility should be well organized and have a plan to address all risks to the quality and safety of the water supply.

Recently, the water industry has begun to move towards a more proactive approach to managing the safety of water supplies by incorporating quality assurance principles. The US drinking water industry employs several quality assurance principles, especially for the control of pathogens. For example, the Surface Water Treatment Rule (SWTR), as amended, sets goals for pathogen occurrence in the finished drinking water, and has regulatory provisions for meeting those goals (US Environmental Protection Agency (EPA) 1998). Provisions include watershed control, source water quality, and treatment performance requirements. Pathogen monitoring is not required under this rule. Additionally, the Total Coliform Rule requires periodic on-site sanitary surveys. The goal of monitoring for the SWTR and other microbial regulations is to assess the effectiveness of pathogen control measures, using monitoring tools such as total coliforms, fecal coliforms, *E. coli*, water turbidity, and disinfectant levels.

The AWWA has several voluntary programs that incorporate quality assurance principles for controlling pathogens in drinking water: QualServe, the Partnership for Safe Water and a new distribution system standard. QualServe is a continuous quality improvement program that helps utilities improve overall service using a self-assessment tool, a peer review process, and a benchmarking clearinghouse. The Partnership for Safe Water, a joint initiative among AWWA, EPA, and other drinking water organizations, also uses a self-assessment tool and a peer review process to optimize treatment plant performance for systems using surface water supplies. AWWA Standard G200-Distribution Systems Operation and Maintenance, effective May 1, 2004, describes critical elements for the operation and management of water distribution systems. This standard is currently being piloted at two utilities along with G100-Standard for Water Treatment Operations and Management and a draft standard, G-300, which covers source water protection (AWWA 2005). The pilot study will evaluate the practicality of these standards as part of a long-range process of standards development that could serve as the basis for a future water utility accreditation program.

This proactive, risk-based management approach is occurring on an international level as emphasized by the World Health Organization (WHO) guidelines (3rd edition) and the complementary 2004 Bonn Charter for Safe Drinking Water that are discussed in Chapter 2. In Australia, the McClellan Inquiry into the 1998 Sydney Water *Cryptosporidium* contamination incident recommended introducing quality assurance procedures as a framework for guiding water quality protection (Davison, Davis, and Deere 1999). A major revision to Australian food legislation in 1999 included tap water in the definition of food and required a quality assurance system incorporating Hazard Analysis Critical Control Point (HACCP) principles for all food suppliers to assure food safety (Davison, Davis, and Deere 1999). More recently, the Walkerton Inquiry in Canada also concluded, "Perhaps the most significant recommendations in this report address the need for quality management through mandatory accreditation and operational planning" (O'Connor 2002). The recommended quality management system should include real time process control and preventive strategies to identify and manage risks to public health (O'Connor 2002).

Quality assurance principles and procedures, such as the HACCP system, are important in controlling risk. The EPA defines quality assurance as "an integrated system of activities involving planning, quality control, quality assessment, reporting and quality improvement to ensure that a product or service meets defined standards of quality with a stated level of confidence" (EPA 1991). The EPA defines quality control as "the overall system of technical activities whose purpose is to measure and control the quality of a product or service so that it meets the needs of users. The aim is to provide quality that is satisfactory, adequate, dependable, and economical" (EPA 1991). Application of quality assurance principles may have some value for water supply since it is very hard to manage the quality control of drinking water between release from storage and the point of consumption.

Quality assurance systems are incorporated into production and service delivery processes across the developed world. There are a number of standards and guidelines available, with International Organization for Standardization (ISO) being the internationally recognized standards that are commonly applied in Europe, Australia, and Asia. Two ISO Standards commonly employed by water utilities include ISO Standards 9001 and 14001. ISO Standard 9001 defines a Quality Management System that demonstrates the ability of an organization to consistently provide products and services that meet customer needs, regulatory requirements and internal goals (Nyman 2004). ISO Standard 14001 defines an Environmental Management System (EMS) that addresses potential impacts on the environment (Nyman 2004).

"HACCP is a tool to assess hazards and establish control systems that focus on prevention rather than relying mainly on end-product testing" (WHO 1997). The application of HACCP in a systematic manner helps the water utility to identify hazards closer to the source of the hazard, thus minimizing the occurrence and effects of incidents that may impact the safety and quality of the water.

A quality assurance system such as HACCP could have helped to prevent the waterborne disease outbreak in Walkerton, Ontario, Canada where a breakthrough of *E. coli* O157:H7 and *Campylobacter* caused seven deaths and more than 2,300 cases of waterborne disease out of a town population of 5,000 (Hrudey and Walker 2005). The tragic event "...severely undermined the trust many Canadians had in their municipal water supplies...and also cost Ontario taxpayers hundreds of millions of dollars...The official inquiry into the events leading to the tragedy showed that if the system had been monitored properly and the operators had responded effectively to the signs of trouble, the severe outbreak would have been prevented or substantially reduced" (Hrudey and Walker 2005).

A HACCP system could also help prevent a loss of confidence in water supply safety from a suspected contamination incident. The Sydney *Cryptosporidium* contamination incident did not cause illness or death, but it affected 3.6 million consumers and had an estimated cost of about $15 million (US Dollars) (Chapman, Jayaratne, and Pamminger 2003), plus resulted in job loss for key utility personnel.

PURPOSE OF PROJECT

Since the mid-1990s, the HACCP system has been applied to a number of drinking water systems in Australia, New Zealand, and several European countries. The experiences of these water systems show that HACCP offers many benefits to a water system. The purpose of this project was to further evaluate the application of HACCP in the drinking water industry to protect and maintain distribution system water quality. A second objective was to prepare guidance for water utilities to encourage broader application of HACCP. These guidelines include information on how to integrate the HACCP approach into existing utility practices.

RESEARCH STRATEGY

The approach used to meet the Project objectives consisted of five tasks:

- Task 1—Develop HACCP Model for Water Distribution Systems
- Task 2—Conduct Peer Review of HACCP Model
- Task 3—Pilot Test HACCP System
- Task 4—Integrate HACCP System With Existing Practices
- Task 5—Prepare Project Deliverables.

In Task 1, a tailored HACCP system was developed specifically for water distribution systems. This tailored plan was primarily based on water industry experiences in Australia applying HACCP to individual water systems. Other key sources of information included the WHO drinking water guidelines, HACCP applications in water-related food industry sectors (aquaculture, bottled water, and packaged ice), and New Zealand water industry and regulatory sector experiences.

In Task 2, the draft HACCP plan for distribution systems was subjected to peer review. Reviewers included water utilities and regulatory agencies with experience of applying HACCP to water supplies. Specifically, the review was undertaken in consultation with:

- The five HACCP-certified participating utilities of the Australian Cooperative Research Center for Water Quality and Treatment (CRCWQT);
- Regulators with experience regulating through HACCP approaches (New Zealand Ministry of Health, South Australian Department of Human Services and Victorian Department of Human Services); and
- Researchers with experience developing HACCP-based systems (Monash Medical School, Melbourne).

In Task 3, the tailored HACCP system was to be pilot tested by four participating utilities including Sydney Water Corporation in Australia, Power and Water Corporation in Australia, the City of Austin, Texas, and the South Berwick Water District in Maine. A training workshop was held at each utility location so that the Project Team could train utility staff on the HACCP approach, and to initiate development of the utility's HACCP plan. Each participating utility formed a HACCP team to further develop the HACCP plan and to guide its implementation. The goal was for each utility to implement their HACCP plan over a 12-month period during which certain operational and water quality parameters would be monitored.

In Task 4, the Project Team used findings from the first three tasks to formulate guidelines on developing and implementing HACCP Plans, and integrating the HACCP approach with current practices. Current practices for distribution system protection include sanitary surveys, water quality monitoring programs, an Emergency Operations Plan, standard operating procedures, ISO 9000 series quality system certification, QualServe activities, and daily operational procedures. Additional guidelines and feedback on HACCP implementation was prepared in the form of utility case studies which captured the experiences of utilities that have operated with certified HACCP systems in place for several years (Chapter 5).

In Task 5, the Project Team prepared periodic reports to document progress and solicit input from both AwwaRF and the Project Advisory Committee (PAC) throughout the course of the project. These periodic reports included interim work products, status reports, responses to PAC comments, and reports on the project team's outreach activities.

USE OF REPORT

This report may be useful to water utilities that are considering whether or not to apply the HACCP system to their distribution system. These readers are referred to the following chapters:

- In Chapter 2, background information on HACCP and current applications are described.
- In Chapter 3, the reader will find guidance on how to implement HACCP including resource requirements, the need for supporting utility programs, and standard operating procedures.
- In Chapter 4, short-term pilot-studies (12 months duration) evaluate HACCP based on six different categories: costs and benefits, regulatory compliance, customer satisfaction, risk management, system management, and human factors.
- In Chapter 5, several case studies from HACCP-certified utilities in Australia evaluate HACCP based on the same criteria as the pilot studies in Chapter 4. These case studies also provide insight into the challenges of implementing HACCP.

For water utilities that are ready to start applying HACCP to their water distribution system, this report presents a tailored HACCP system that can be used as a starting point for any distribution system. The reader is referred to Chapter 3 where detailed step-by-step instructions explain how to develop a HACCP plan. Example workshop training materials are provided in Appendix A. Example HACCP Plans for the four participating utilities are provided in Appendix B. Pilot study reports can be found in Appendix C. Appendix D contains Management Plans Sydney Water developed as part of the utility HACCP process.

For industry researchers that are interested in furthering the knowledge and applications potential for HACCP in drinking water systems, research needs are outlined in Chapter 6.

WATER QUALITY FRAMEWORK

It is important to recognize that water utilities operate within a very broad regulatory and operational framework that collectively contributes to, and controls, the safety and quality of drinking water produced. The term "Water Quality Framework" is used here to refer to this overall framework within which water safety/quality is both regulated by the government and managed by utilities and their contractors. Within this Framework is a Water Quality Management Plan that is specifically concerned with the operational management by the water utility and its contractors. The Water Quality Management Plan serves as a road map, linking the HACCP Plan with existing supporting programs and demonstrating that there has been a cross-checking and augmentation of existing organizational risk and quality management systems. The interrelationships of these elements within the overall Framework are illustrated in Figure 1.1.

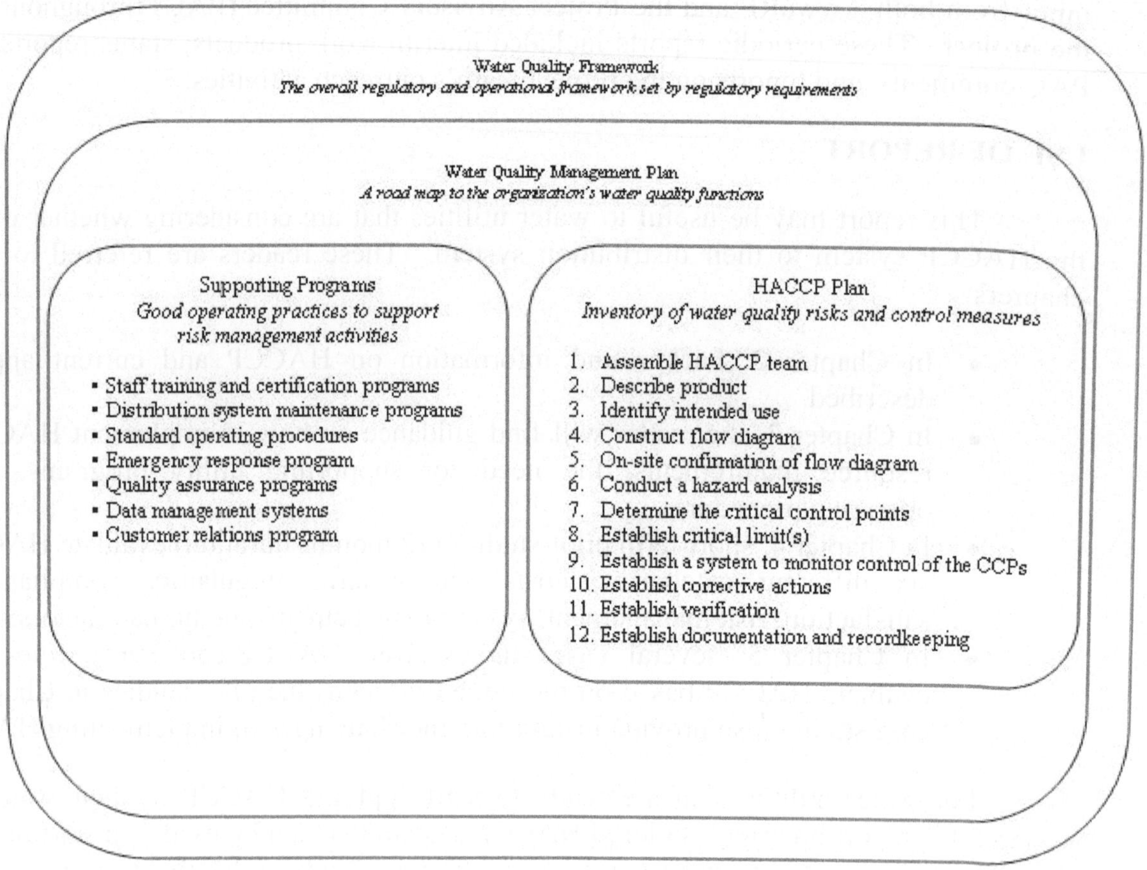

CCPs = critical control points

Figure 1.1 Illustration of HACCP Plan as integral part of Water Quality Framework

CHAPTER 2
INTRODUCTION TO HACCP

BACKGROUND

In 1959, the National Aeronautics and Space Administration (NASA) was planning for manned space missions and was concerned about possible contamination of the astronauts' food from "...potentially catastrophic disease-producing bacteria and toxin" (NASA 1991). To solve this problem, NASA enlisted the help of the Pillsbury Company to develop the HACCP concept "...to establish control over the entire process, the raw materials, the processing environment and the people involved" (NASA 1991). The Pillsbury Company developed the basic HACCP concepts with cooperation and participation from NASA, the Natick Laboratories of the US Army, and the US Air Force Space Laboratory Project Group (Mucklow 1997).

Since the 1980's, the HACCP system has been widely adopted by the food and beverage industries world-wide where it forms an important part of their "food safety plans." Quality assurance systems incorporating HACCP principles have become the benchmark means to assure food and beverage safety since its codification in 1993 by the United Nations Food and Agricultural Organization (FAO) and WHO (Deere and Davison 1999). In 1996, the US Food and Drug Administration (FDA) and the US Department of Agriculture Food Safety and Inspection Service (FSIS) promulgated regulations requiring the use of HACCP in the seafood, red meat, and poultry industries. The WHO guidelines for HACCP, known as Codex Alimentarius, have been adopted internationally as the primary recognized food safety methodology for risk management. The current HACCP guideline was developed in 1997 by the Codex Alimentarius Commission (http://www.who.int/foodsafety/codex/en), an inter-governmental body established to implement the Joint FAO/WHO Food Standards Program.

By helping to improve food production processes to prevent contamination, the HACCP system can reduce or prevent the occurrence of food borne illnesses. In October 2003, the FDA/FSIS reported four consecutive annual drops in human Listeria infection and a 70 percent decline in positive food samples compared with years prior to HACCP implementation (Fok and Emde 2004).

The bottled water industry employs additional measures beyond regulatory requirements to help ensure the safety and quality of bottled water beginning with the source through packaging. Bottler members of the International Bottled Water Association (IBWA) must adhere to the IBWA Model Code (available at www.bottledwater.org), which requires members to develop a HACCP program for each of their facilities. The HACCP program requirement further ensures food safety and security within the production facility. The IBWA Model Code is, in several cases, more stringent than state and federal regulations and has been adopted by more than a dozen states as their standard for regulation of bottled water. At the federal level, bottled water is regulated as a packaged food product, governed by the FDA through the Food, Drug, and Cosmetic Act. At the state level, bottled water is regulated in myriad ways, typically through state environmental, food, or agricultural agencies.

THE HACCP SYSTEM

The Codex Alimentarius Commission defines 12 sequential steps for planning and implementing a HACCP system. The information prepared in completing these 12 steps constitutes the utility's HACCP Plan. These steps are summarized below:

Step 1 – Assemble HACCP Team - Pull together a multidisciplinary team to prepare, develop, verify, and implement the plan.
Step 2 – Describe Drinking Water - Describe the utility's drinking water, including its source, treatment, storage, distribution and any existing standards for quality and safety.
Step 3 – Identify Intended Use - Describe how the drinking water is used and the major users.
Step 4 – Construct Flow Diagram - For a comprehensive HACCP plan, this would be a schematic showing sources of water, details of treatment, storage, pumping, and distribution to end users. For a HACCP plan directed towards a distribution system, the schematic would be restricted to showing the water flow path from the treatment plant to end users.
Step 5 –Confirm Flow Diagram – Since the flow diagram is a critical element used as a basis for the HACCP plan, its accuracy should be confirmed by the HACCP team.
Step 6 - Conduct a Hazard Analysis - Using the process flow diagram, identify hazards, their likelihood of occurrence, potential consequences, and control measures.
Step 7 – Determine the Critical Control Points (CCPs) - For each significant hazard, identify points in the process where the consequences of failure are irreversible.
Step 8 – Establish Critical Limit(s) - Determine critical limits for the CCPs that will trigger a corrective action. A critical limit is a criterion which separates acceptability from unacceptability.
Step 9 – Establish a System to Monitor Control of the CCPs - Establish monitoring points, frequency, and responsibility.
Step 10 – Establish Corrective Actions - Develop plans for follow-up activity when critical limits are exceeded.
Step 11 – Validate/Verify HACCP Plan - Have the HACCP team and other affected parties check the HACCP plan for accuracy, ability to implement, and potential effectiveness.
Step 12 – Establish Documentation and Recordkeeping - Develop a record keeping system to track system performance at CCPs.

These steps of HACCP are discussed in more detail in Chapter 3 as applied to a tailored HACCP plan for water distribution systems.

HACCP IN DRINKING WATER REGULATIONS AND GUIDELINES

In 2004, the WHO issued the 3rd edition of its drinking water guidelines that outline a framework for drinking water safety based on the multiple barrier approach and several risk management and quality management approaches including the HACCP principles (Davison and Bartram 2004). The WHO framework has three main components: health-based water quality targets based on public health protection and disease prevention; a Water Safety Plan as described below; and independent surveillance activities including audits of the Water Safety

Plan and final checks on the finished drinking water. WHO recommends that water suppliers develop a Water Safety Plan that documents the following major elements:

1. A source-to-tap system assessment that determines whether a water system can deliver water meeting certain water quality targets;
2. Control measures for identified hazards and operational monitoring of control measures; and
3. A management plan that documents the system assessment, control measures, the monitoring plan, corrective action procedures to address water quality incidents, communication plan, and supporting programs such as standard operating procedures (SOPs), employee training, and risk communication.

The 2004 Bonn Charter for Safe Drinking Water complements the WHO guidelines in providing international guidance on drinking water quality management (Breach 2004). This Charter is the end-product of two expert workshops held in Bonn, Germany in October 2001 and February 2004, and is applicable to all water systems worldwide. The Charter's key principles include the following (Breach 2004):

1. Good, safe drinking water can only be provided reliably and consistently through an integrated, source-to-tap approach.
2. The integrated approach requires close cooperation and partnerships among governments, water suppliers, health agencies, environmental agencies, land users, contractors, plumbers, consumers, and manufacturers of materials, products, and devices used in the water supply system (Breach 2004).
3. Traditional verification of drinking water quality based on measurement of various parameters against predetermined standards or guidelines will continue to play a critical role in ensuring drinking water quality. In the future, there should be much greater emphasis on use of preventative, risk-based management control systems.
4. The quality assurance process and derivation of specific standards need to be transparent to assure consumer confidence.
5. The standards used to measure water quality and safety can legitimately vary between different countries and regions depending on local circumstances. However, these standards should ensure the provision of water that has the trust of consumers, is affordable, and meets the following minimum criteria (Breach 2004):
   a) Does not pose a health threat to consumers;
   b) Is acceptable to consumers in terms of taste, odor and appearance; and
   c) Is reliable in terms of both quality and quantity.
6. Effective approaches to managing drinking water quality rely on an interlinked set of processes which must involve the following three elements (Breach 2004):
   a) Establishment of clear responsibilities and institutional arrangements for the different stakeholders.
   b) Implementation of effective control systems directed to identifying and managing risks, thereby mitigating their impacts (Drinking Water Quality Management Plans).
   c) Assessment of compliance against the necessary minimum standards for drinking water quality (verification).

7. A Drinking Water Quality Management Plan that includes the following three elements (Breach 2004):
   a) A system-wide, risk-based assessment of safety from source-to-tap;
   b) Identification of the most effective control points to reduce the risk; and
   c) Effective systems and operational plans to deal with both routine and abnormal operating conditions.

Countries which use WHO guidelines as the minimum criteria for water system regulation will need to recommend or require that water utilities develop Water Safety Plans. In several countries, HACCP principles are being incorporated into national guidelines and standards as summarized in Table 2.1.

Table 2.1
HACCP programs for controlling public health risks in drinking water

Country/ Organization	Regulation or Regulatory Guideline	Website for current information	Reference
WHO	Guidelines for Drinking Water Quality (3rd Edition)	www.who.int/water_sanitation_health/dwq/guidelines2/en/	Davison and Bartram (2004)
Australia	Framework for Management of Drinking Water Quality (Guideline)	www.nhmrc.gov.au/publications/synopses/eh19syn.htm	Cunliffe (2004)
New Zealand	Public Health Risk Management Plans (Proposed regulation)	www.moh.govt.nz/water Select "publications" then "Public Health Risk Management Plans"	Nokes (2001)
Switzerland	W1002 Regulatory guideline: Recommendations for a Simple Quality Assurance System for Water Supplies (Guideline)	www.svgw.ch (German, French and Italian only)	Swiss Gas and Water Industry Association (2003)
	Hygiene Ordinance (SR 817.051, HyV), Article 11 (Regulation)	www.svgw.ch (German, French and Italian only)	Bosshart (2003)
France	French National Transcription: Decree 2001-1220 (Dec. 20, 2001) Water Safety for Human Health, Risk Assessment and Management; Article 18-2 (Regulation)	None found	Metge (2003); DeBier and Joret (2004)
Ontario, Canada	Province-wide initiative; Water quality management system based on HACCP, ISO 9001, and ISO14001 (Guideline)	None found	Kuslikis and White (2004)

The 2004 revision of the Australian Drinking Water Guidelines includes the "Framework for Management of Drinking Water Quality" which is intended to provide guidance on establishing preventive, source-to-tap risk management systems for drinking water quality (National Health and Medical Research Council and Natural Resource Management Ministerial Council 2004). These guidelines are not mandatory or legally enforceable standards. In Australia, drinking water is regulated by State health agencies. In 2003, the State of Victoria (Australia) introduced legislation, the Safe Drinking Water Act of 2003, requiring that all water utilities implement a risk management plan based on the Framework (Cunliffe 2004). The requirements of this Act are primarily addressed by implementing a HACCP Plan.

The Australian Framework has been designed to reduce the reliance on compliance monitoring as the primary means for managing water quality, and instead promotes monitoring as a verification tool for assuring that preventive measures are working effectively. "The Framework was derived by supplementing the information on preventive system management already provided in the Australian Drinking Water Guidelines with principles described in existing quality management systems such as ISO 9001 (Quality Systems), ISO 14001 (Environmental Management Systems), Australian and New Zealand Standard (AS/NZS) 4360 (Risk Management) and the HACCP system" (Cunliffe 2001) The Framework includes twelve main elements (Cunliffe 2004):

1. Commitment to Drinking Water Quality Management;
2. Assessment of the Drinking Water Supply System;
3. Planning-Preventive Strategies for Drinking Water Quality Management;
4. Implementation-Operational Procedures and Process Control;
5. Verification of Drinking Water Quality;
6. Incident and Emergency Response;
7. Employee Awareness and Training;
8. Community Involvement and Awareness;
9. Research and Development;
10. Documentation and Reporting;
11. Evaluation and Audit; and
12. Review and Continual Improvement.

In New Zealand, new legislation amended the Health Act of 1956 to require water suppliers to develop a so-called "Public Health Risk Management Plan" (Nokes 2001). To assist water suppliers in developing and implementing this Plan, the Ministry of Health has prepared a series of guides that are based on the risk management framework contained in AS/NZS 4360:1999 (Standards Australia/Standards New Zealand 2004) and HACCP methodology (Codex Alimentarius Commission 1993). These guides contain the following information:

- Potential problems during different processes and operations that might allow contaminant intrusion;
- Corrective actions when contamination occurs; and
- Preventive measures to reduce the likelihood of the problems recurring.

A separate booklet discusses the overall approach to developing and implementing a Plan as outlined in Table 2.2 (Ministry of Health 2001).

Table 2.2
New Zealand approach for preparing Public Health Risk Management Plans

Step No.	Description	Contribution to the Public Health Risk Management Plan
1	Identify the elements in the supply and the PHRMP* Guidelines needed.	Flow diagram of the supply
2	Identify which barriers to contamination are present.	Check-list of barriers present in the supply
3	Identify events that may introduce hazards into the water.	
4	Use the Guides to identify: • Causes • Preventive Measures • Corrective Actions	Risk information table for the supply overall
5	Decide where improvements should be made	
6	Decide on the order in which improvements need to be made	Improvement Schedule • Improvements needed • Levels of importance • Timetable • Responsibilities
7	Draw up a timetable for making the improvements	
8	Identify links to other quality assurance systems	Note of other quality assurance systems
9	Prepare contingency plans	Contingency plans for each supply element
10	Prepare instructions for Performance Assessment of the Plan	Set of instructions for review of the performance of the PHRMP*
11	Decide on communication policy and needs	Set of instructions for reporting

Source: Ministry of Health (2001)
*PHRMP – Public Health Risk Management Plan

In Switzerland, Article 11 of the hygiene regulation (SR 817.051, HyV), requires application of the HACCP principles. A regulatory guideline (W1002) entitled "Recommendations for a Simple Quality Assurance System for Water Supplies" has been prepared to assist water utilities in complying with this requirement. The regulatory guideline recommends the following nine-step approach (Swiss Gas and Water Industry Association 2003):

1. Organization, Responsibilities, and Expertise;
2. Survey of the Water Supply (develop process flow diagram);
3. Assessment of the Water Supply (evaluate hazards, list critical points);
4. Elimination of Hazards (Critical Points);

5. Reduction of Hazards by Maintenance;
6. Control of Hazards (Critical Points);
7. Practical Implementation of the Instructions;
8. Annual Evaluation of the Water Supply; and
9. Confirmation of Self-Assessment by Third Party.

In Francc, Article 18-2, Optimization of Monitoring, of the French National Transcription: Decree 2001-1220 (Dec. 20, 2001) entitled Water Safety for Human Health, Risk Assessment and Management requires risk assessment, identification of CCPs and control measures.

In Ontario, Canada, the Ontario Ministry of the Environment is currently developing a Water Quality Management Standard as a result of the Walkerton tragedy and the follow-up investigation. A province-wide initiative has resulted in the development of an integrated risk management system based on HACCP, ISO 9001, and ISO 14001 (Kuslikis and White 2004). The system is being employed in a number of cities and municipalities at water treatment facilities, wastewater treatment facilities, solid waste facilities, and engineering facilities. Several reasons given for employing such an initiative include (Kuslikis and White 2004):

- To promote water quality excellence to the public;
- To secure knowledge potentially lost with retiring employees;
- To satisfy a need for consistent application of best practices and continual improvement; and
- To supplement end product failure testing with a quality assurance system/quality control system.

HACCP APPLICATIONS IN THE WATER INDUSTRY

Bryan (1993) and Havelaar (1994) first introduced the idea of applying HACCP to drinking water systems. Bryan (1993) presented the HACCP approach as a way to improve water treatment processes to reduce the occurrence of waterborne disease. He also noted the need to address distribution system inadequacies (e.g. ingress from contaminated surface water or sewage if the distribution system is poorly maintained) that could affect the quality of finished water. However, Bryan (1993) cautioned that the use of HACCP for the water industry was limited for a number of reasons:

1. The structure, equipment and cleaning standards may be inappropriate.
2. Effective communication may be lacking, preventing swift action when a problem occurs.
3. The appropriate corrective actions may not be clearly documented.
4. The causative agent of waterborne disease outbreaks cannot always be isolated from either the water supply or the human case due to the lack of analytical methods for many pathogenic viruses and other microorganisms associated with drinking water samples. Most existing analytical methods do not utilize on-line technologies, preventing an instantaneous reaction to failure. This can be addressed by monitoring physical parameters to control critical points (e.g. monitor particle counts to assess the presence of Cryptosporidium oocysts).

5. If HACCP is applied only to treatment facilities and not the distribution system, it may not prevent a waterborne disease outbreak caused by distribution system inadequacies.

Havelaar (1994) examined the applicability of HACCP to drinking water supply with a focus on microbiological contaminants. He introduced a generalized HACCP analysis for drinking water production, including source, treatment and distribution process steps, listing typical hazards, preventive measures, CCPs, monitoring procedures, and corrective actions. Havelaar (1994) noted that a key issue is properly identifying the CCPs (i.e. those points within the system or its operation whose disruption or failure would result in a greater public health risk compared to other points) because the major efforts in process control will be directed towards these steps. In many food operations, the heating step or another single step is the major barrier to pathogens. In water systems, multiple steps should be considered as CCPs because a multiple barrier approach is used to control microorganisms (e.g. source protection, filtration, and disinfection).

Since the mid-1990's, HACCP has been applied by a number of individual drinking water systems. Several water utilities in Australia have independently audited, certified HACCP systems: South East Water, Yarra Valley Water, Melbourne Water, Brisbane City Council, and Gold Coast Water. In Iceland, Reykjavik Energy has been utilizing an accredited HACCP system since 1997 for its water works (Gissurarson 2004). In France, Veolia Water has implemented a new comprehensive management system based on HACCP at 40 different water production/water distribution systems (DeBier and Joret 2004). In Ontario, Canada, several municipalities and cities are developing and implementing an integrated risk management system based on ISO 9001, ISO 14001, and HACCP (Kuslikis and White 2004). Smaller systems have applied HACCP principles too, such as the councils in the Australian State of New South Wales and small supplies in New Zealand.

Watershed Protection

The HACCP system may be used to improve how a utility's watershed management program is structured. For example, a successful HAACP Plan was implemented by San Francisco to address the concern that livestock may be a primary source of Cryptosporidium parvum (Barry et al. 2004). The utility was looking for a solution focused on preventive measures and one that provided a systematic approach that could be duplicated throughout the watershed. The HACCP Plan identified five management areas to control waterborne pathogens in the watershed – feral pig management, livestock/grazing, wildlife management, human/ recreation management, and treatment plant management. For each management area (except the treatment plant), the Plan identified CCPs, management measures, monitoring, critical limits, corrective actions, and record-keeping needs. San Francisco's HACCP Plan was also expanded to control physical hazards in the watershed, such as sediment.

Hazards in the watershed may stem from recreational activities, urban storm water systems, animal feed lots, and/or licensed discharges, such as discharges from sewage treatment plants (Deere 2004). Hazards may be considered in two separate categories – those that will be significantly reduced by treatment downstream (e.g. bacterial pathogens if disinfection is practiced) and those that will not be significantly reduced by treatment (e.g. pesticides if ozone/carbon treatment is not utilized) (Deere 2004).

The HACCP system may improve existing watershed management programs through its emphasis on feedback loops in monitoring, corrective actions, and documentation. Monitoring activities in watershed management programs include inspections as well as water quality testing. For example, a control measure for the livestock "hazard" may be to install fencing to keep animals away from the water supply reservoir. To determine if this control measure is effective, the utility may opt to inspect the fencing on a weekly basis to make sure it is intact. When an inspection is not completed according to the frequency specified in the HACCP Plan, or when the inspection identifies problems, corrective actions must be taken (as specified in the Plan). The HACCP Plan includes provisions for documenting these corrective actions and follow-up monitoring or inspection activities to assure that the problems are followed through to a resolution.

Havelaar (1994) presents a generalized application of HACCP to water system processes associated with the source of supply, as summarized in Table 2.3. For each process step (e.g. groundwater abstraction), typical hazards, preventive measures, monitoring procedures, and corrective actions are summarized.

Table 2.3
Generalized HACCP analysis of source of supply

Hazards	Preventive measures	Critical control point (CCP)?	CCP parameters	Monitoring procedures	Corrective actions
Process step – groundwater abstraction					
Transport of pathogens to wellhead	Define protection zone around well and restrict land use	Yes	Traveling time	Tracer injection studies, specific pathogens, fecal index bacteria	Remove sources of pollution
Ingress of pathogens through well casing	Proper construction and maintenance	Yes	Adhere to good engineering practices	Inspection, fecal index bacteria	Instruction/re-construction
Process step – bank infiltration					
Transport of pathogens to wellhead	Define minimum traveling time and/or distance	Yes	Site-specific	Tracer injection studies, specific pathogens, fecal index bacteria	Replace abstraction wells, increase treatment
Ingress of pathogens through well casing	Proper construction and maintenance	Yes	Adhere to good engineering practices	Inspection, fecal index bacteria	Instruction/re-construction

(Continued on next page)

Table 2.3 (Continued)
Generalized HACCP analysis of source of supply

Hazards	Preventive measures	Critical control point (CCP)?	CCP parameters	Monitoring procedures	Corrective actions
Process step – surface water abstraction					
Contamination by fecal discharges	Reduce point and diffuse pollution sources	No		Fecal index bacteria, specific pathogens, turbidity	Increase treatment
Multiplication of pathogens	Control eutrophication, thermal discharges, residence time of water	No		Not applicable	
Process step – storage of surface water in reservoirs					
Short circuiting	Build reservoirs in series	No		Tracer studies, conservative parameters, fecal index bacteria	Increase treatment
Recontamination by feces of wildlife	Discourage presence of wildlife	No		Specific pathogens	

Source: Adapted from Havelaar (1994). Reprinted from Food Control, Vol. 5 No. 3, Havelaar, Application of HACCP to Drinking Water Supply, pp.145-152, 1994, with permission from Elsevier.

Treatment Facilities

Smith (2004) found that implementing HACCP for treatment systems at Gold Coast Water in Australia was a "...relatively simple matter." Fok and Emde (2004) state that water treatment plants are ideally suited for implementing HACCP because these facilities typically employ a single process train with a single end-product. Typical hazards, preventive measures, monitoring procedures, and corrective actions are summarized in Table 2.4 (Havelaar 1994).

Table 2.4
Generalized HACCP analysis of treatment process

Hazards	Preventive measures	Critical control point (CCP)?	CCP parameters	Monitoring procedures	Corrective actions
Process steps – pretreatment, coagulation/flocculation/sedimentation/filtration					
Poor floc formation, poor floc removal, filter defects	Increase coagulant dose, add coagulant aid, regular backwashing and cleaning, first filtrate after backwash to waste	Yes	Turbidity, particle counts, pressure loss	On-line monitoring	Increase disinfection
Process step – disinfection					
Survival of pathogens	Optimize dose and contact time of disinfectant	Yes	Residual concentration of disinfectant (may vary during the year), pH, temperature, bacteriological indicator organisms	On-line monitoring	Automatic feedback system
Formation of disinfection by-products	Optimize dose and contact time of disinfectant	Yes	Residual concentration of disinfectant (may vary during the year), pH, temperature		Modify target dose/residual

Source: Adapted from Havelaar (1994). Reprinted from Food Control, Vol. 5 No. 3, Havelaar, Application of HACCP to Drinking Water Supply, pp.145-152, 1994, with permission from Elsevier.

Water Distribution Systems

The HACCP model may be applied to water distribution systems to prioritize CCPs and focus the system operator's resources on these locations and processes (Sobsey et al. 1993; Bryan 1993; Deere and Davison 1999). Typical hazards, preventive measures, monitoring procedures, and corrective actions are summarized in Table 2.5 (Havelaar 1994).

Table 2.5
Generalized HACCP analysis of water distribution systems

Hazards	Preventive measures	Critical control point (CCP)?	CCP parameters	Monitoring procedures	Corrective actions
Recontamination from cross-connections and storage facilities	Adequate construction; positive pressure at all times	Yes	Total coliform bacteria; system pressure; disinfectant residual	Frequent samples; testing of backflow prevention devices	Isolate part of system; rechlorination
Contamination at repair and construction sites	Sanitary practices during construction	Yes	Adhere to sanitary practices	Inspection, sample	Flushing, disinfection, worker training, program assessment
Regrowth of opportunistic pathogens	Reduce residence time; reduce assimilable organic carbon and/or biofilm potential	Po ssibly, system dependent	Disinfectant residual, total coliform bacteria, assimilable organic carbon, water temperature	Frequent monitoring	Flushing, disinfection, treatment optimization, reduce water age

Source: Adapted from Havelaar (1994). Reprinted from Food Control, Vol. 5 No. 3, Havelaar, Application of HACCP to Drinking Water Supply, pp.145-152, 1994, with permission from Elsevier.

Other Applications in the Water Industry

The Capitol Health region in Edmonton, Alberta, Canada used HACCP principles to develop a new boil water advisory protocol in 1998 (Fok and Emde 2004). “The use of the HACCP process resulted in a better understanding of monitoring parameters and fostered communication and understanding between…[the health department and the water utility]” (Fok and Emde 2004).

SUMMARY

The HACCP concept was first developed in 1959 by the Pillsbury Company to improve food safety for NASA’s manned space missions. Since the 1980’s, the HACCP system has been widely adopted by the food and beverage industries world-wide where it forms an important part of their “food safety plans.” In the US, certain food production establishments such as seafood, red meat, and poultry industries are required by federal regulations to develop and implement a HACCP plan. The IBWA requires its members to develop a HACCP program for each of their facilities.

The current HACCP guideline, developed in 1997, by the Codex Alimentarius Commission includes 12 sequential steps for planning and implementing a HACCP system. The information prepared in completing these 12 steps constitutes the utility's HACCP Plan.

HACCP has been incorporated in many drinking water regulatory requirements and guidelines around the globe. In 2004, the WHO issued the 3rd edition of its drinking water guidelines that outline a framework for drinking water safety based on the multiple barrier approach and several risk management and quality management approaches including the HACCP principles (Davison and Bartram 2004). Countries that use WHO guidelines as the minimum criteria for water system regulation will need to recommend or require that water utilities develop Water Safety Plans. In Switzerland, Article 11 of the hygiene regulation (SR 817.051, HyV), requires application of the HACCP principles. In New Zealand, new legislation amended the Health Act of 1956 to require water suppliers to develop a so-called "Public Health Risk Management Plan" (Nokes 2001).

Bryan (1993) and Havelaar (1994) first introduced the idea of applying HACCP to drinking water systems. Havelaar (1994) introduced a generalized HACCP analysis for drinking water production, including source, treatment and distribution process steps. This generalized analysis, focused on microbiological contaminants, listed typical hazards, preventive measures, CCPs, monitoring procedures and corrective actions, as summarized in Tables 2.3 through 2.5. Since the mid-1990's, HACCP has been applied by a number of individual drinking water systems in Australia, France, Canada, and New Zealand, amongst others.

CHAPTER 3
TAILORED HACCP GUIDE FOR WATER DISTRIBUTION SYSTEMS

INTRODUCTION

In Chapter 2, HACCP terminology and the 12 sequential steps for planning and implementing a HACCP system were introduced. The information prepared in completing these 12 steps constitutes the utility's HACCP Plan.

The purpose of Chapter 3 is to provide a tailored HACCP manual to guide the production of a distribution system HACCP plan by a water utility. Although a utility could independently produce and implement a HACCP plan based on this report, training and coaching from people experienced in the application of HACCP to water is likely to greatly simplify this task.

Prior to initiating a HACCP Plan, a utility should complete the following tasks:

- Identify goals and expected benefits;
- Estimate resource requirements;
- Solicit employee support and commitment;
- Conduct a HACCP training workshop; and
- Identify and update supporting programs and practices.

These activities are discussed in the following introductory sections.

Identify Goals and Expected Benefits

Prior to initiating a HACCP Plan, it is important for the utility to decide on its overall goals for the program. This helps to focus resources most efficiently and to help sell the program to utility management, staff, and customers. The HACCP system may provide both tangible and intangible benefits to water utilities, including (Fok and Emde 2004; Davison and Deere 2004; Smith 2004; and Kuslikis and White 2004):

- Reduction in customer complaints;
- Improved water quality;
- Improved work processes, such as SOPs, monitoring strategies, documentation procedures, and communication methods;
- Improved understanding of risks and risk management;
- Improved employee skills; and
- Demonstration of due diligence.

Specific case study examples of these benefits are described in Chapters 4 and 5. For example, the City of Austin, Texas found that their HACCP pilot study helped improve employee skills and knowledge including:

- The need to respond quickly to main breaks in small pressure zones;
- The location of pressure zone boundaries;
- The possible occurrence of boundary zone violations;

- The possibility of pressure transients causing low or negative pressure in the distribution system;
- The need to maintain positive pressure at all times; and
- The types of existing data sources.

Estimate Resource Requirements

The utility should estimate the resource requirements for implementing HACCP in their system, and make a firm commitment to providing these resources prior to initiating the process. This section summarizes resource needs identified by utilities that have implemented HACCP plans.

Gold Coast Water in Australia identified the need for a HACCP "champion" who is solely dedicated to administering the HACCP system. "The HACCP champion should have a background in chemistry and microbiology, plus experience in water industry processes" (Smith 2004). Gold Coast Water has not added any other staff to implement HACCP but has purchased additional instrumentation and increased water quality testing. Smith (2004) estimates a 4- to 6-week timeline for developing and implementing a HACCP system for a catchment, treatment, or distribution system with provided there are no existing fundamental flaws. However, the cultural change takes place gradually, over a longer period. The cultural change could take as long as 6 months (Smith 2004a). Other HACCP-certified water utilities in Australia also found that HACCP implementation required a full-time, dedicated coordinator or champion, for at least one month and up to twelve months, as summarized in Table 5.2.

Based on their experiences implementing HACCP in 45 water systems in France, DeBier and Joret (2004) estimate an average of 30 to 60 person-days to implement HACCP in one system (from the catchment to the customers' taps).

Although staff will not be working full time on the HACCP plan for the whole implementation period, (with the possible exception of the HACCP Coordinator), key staff will need to be available to attend workshops, to discuss related policies/procedures/risk management issues, and possibly to develop improved risk management practices.

Some nominal allocation of resources is needed to cover issues that arise during HACCP implementation. For example, some risk assessment conclusions may be uncertain, requiring special investigations, professional services, laboratory testing, and/or procurement of new monitoring equipment; or standard operating procedures may need to be improved to reduce risks to acceptable levels. For example:

- ActewAGL in Canberra, Australia, identified that some of the spot-dosing chemicals being placed into service reservoirs did not have written statements saying they were intended for potable water use. An improvement action was raised whereby materials safety data sheets were obtained from the supplier to confirm that the chemicals were safe and intended for use in potable water systems. (Note: Actew AGL is owned by ACTEW, a government-owned corporation, and AGL, a private company. ACTEW is the asset owner and Actew AGL is the water supply system operator.)
- South East Water in Victoria, Australia, identified that it did not disinfect water mains after repairs, whereas some water authorities did. A special investigation was initiated to find out whether or not post-repair disinfection was required.

Importantly, the resource commitment needs to be made beyond the completion and implementation of the HACCP plan. Once in place, there is a need to occasionally update the plan as well as to maintain the organization's focus on communicating the contents of the plan and complying with it in practice.

As discussed in Chapter 5, the resource needs for five HACCP-certified water utilities in Australia varied from $0 to $55,000 (US Dollars) for HACCP training, coaching, and technical advice, and from $0 to $84,000 (US Dollars) for annual costs, such as water quality monitoring and instrumentation.

Solicit Employee Support and Commitment

To foster employee empowerment and acceptance of HACCP, it is equally important to involve as many employees as possible, including staff with all levels of seniority. The regional municipality of Durham in Ontario, Canada recommends the following strategies to solicit employee support and commitment to HACCP (Kuslukis and White 2004): management support at all levels; top management commitment and participation; and involvement of all staff throughout the process.

Gold Coast Water has observed a progressive change in the organizational culture since implementing HACCP (Smith 2004). Operational staff members were consulted during development of the HACCP plan to identify realistic critical limits and monitoring procedures. Improved monitoring and reporting procedures implemented as part of the HACCP Plan have motivated operational staff to work harder to avoid failures (Smith 2004). Distribution system staff members now consider water quality impacts as part of the many activities they carry out. There is a high level of awareness that accountability is unavoidable and this drives ownership and involvement.

Senior management is intimately involved with the HACCP system at Gold Coast Water because all HACCP "excursions" (water quality or other incidents where critical limits are exceeded) are reported to managers for their information. Management can quickly review these excursion notices and decide if they represent minor incidents or if further inquiry is warranted. In the past, HACCP excursion reports have highlighted problems with staffing, training, asset condition, and system design.

Conduct a HACCP Training Workshop

Where HACCP is most widely used, in the food industry, it is normal practice to begin HACCP implementation by having a significant cross-section of utility staff formally trained in HACCP terminology and methodology. For the case studies referred to in this report, one- or two-day HACCP training courses were undertaken by utility staff. Example training workshop materials are provided in Appendix A. In addition, coaching was provided by people experienced in the practical application of HACCP to water distribution systems during the planning and implementation process. Yarra Valley Water, in Victoria, Australia, undertakes refresher HACCP awareness training each year and includes both contractor and utility staff in these sessions.

Identify and Update Supporting Programs and Practices

Water utilities have many different on-going programs and practices for assessing and correcting deficiencies and risks associated with drinking water supply. Such programs and practices may include:

- Vulnerability assessments;
- Sanitary surveys;
- Emergency response program; and
- Compliance with AWWA distribution system standard.

Common features of these programs include (Cotruvo 2004): risk avoidance and recovery, comprehensive assessment, critical limits/targets, monitoring, plans, recordkeeping, verification/audit, and update/improvement. Some programs emphasize different elements (Cotruvo 2004). For example, the sanitary survey focuses on regulatory and sanitary deficiencies. Vulnerability assessments are primarily concerned with physical security threats.

Prior to initiating a HACCP Plan, it is important to identify these other programs that may contribute valuable information related to the condition of the distribution system and distribution system water quality. The distribution system HACCP plan may be integrated with these programs to avoid duplicative work. The programs may already exist in some form but may need improvement or augmentation. The AWWA Standard G200-Distribution Systems Operation and Maintenance provide specifications for distribution system programs and practices to maintain water quality. Compliance with this standard is recommended prior to implementing HACCP.

Many HACCP registrars (external auditors that provide third or second party certification to attest to the veracity of a HACCP plan) refer to these programs as "Prerequisite Programs" to indicate that their presence is in fact essential. In this report, we have used the term "Supporting Programs" rather than "Prerequisite Programs," but the two are considered synonymous. In Table 3.1, examples are given based on WHO (1997) as well as programs identified on registrar checklists from organizations such as SAI Global, NATA Certification Services International, National Sanitation Foundation (NSF) International, and Lloyds Register.

Organizations that provide HACCP certification services will use the Codex approach as part of their certification criteria. However, different certification authorities and jurisdictions may require slightly different additional Supporting Programs to be in place. Therefore, a utility seeking certification for its HACCP system should check with their prospective registrar(s) at an early stage to find out what the requirements are in their jurisdiction.

Table 3.1
Examples of supporting programs

Program	Description	Example
Management Programs		
Organizational commitment	Staff from all levels of the organization commit to creating and implementing the HACCP plan and to the goals of reliably producing safe, quality water. Senior managerial staff, line supervisors, operations staff, and important contract workers need to collectively buy into this commitment.	Power and Water Corporation, Northern Territory, Australia, has a clear statement of commitment in its HACCP plan, signed by the General Manager for Water. Within the body of the HACCP plan the roles of key individuals in supporting and maintaining the HACCP plan are identified.
Compliance with regulations and guidelines	To help the organization consistently provide reliable, safe, quality water there needs to be a firmly reinforced organizational culture of complying with good operational practices and an intolerance of taking short cuts that lead to work not being performed in a sound and sanitary manner.	Utilities conduct water quality monitoring in the distribution system to comply with Federal Safe Drinking Water Act and State drinking water requirements. Regulations specify where monitoring must be performed, how often, and what laboratory analytical methods must be used.
Sanitary survey	The sanitary survey is an external audit periodically conducted by an independent, qualified third party (e.g. a state primacy agency) to review a water system's effectiveness in producing and distributing safe drinking water.	A sanitary survey of the drinking water distribution system of the District of Columbia was completed in 1995 to identify problems associated with bacteriological activity and to develop recommendations (International Studies and Training Institute, Inc. 1995). The survey found that the distribution system reservoirs had not been drained and inspected in approximately 20 years or more. Survey findings contained 185 recommendations addressing treatment practices, cross connections, distribution system O&M practices, consecutive systems, and operating procedures. Follow-up sanitary surveys were conducted in 1996, 1998 and 2002 to document progress on the original recommendations from the 1995 survey, and to identify any additional needs.
AWWA Distribution System Standards	AWWA Standard G200-Distribution Systems Operation and Maintenance (2004) describes critical elements for the operation and management of water distribution systems. Standard G200 emphasizes the need for specific water quality goals, action plans to respond to problems, and written SOPs.	Water quality requirements of the G200 Standard include: compliance with regulatory requirements; monitoring and control; disinfection residual maintenance; internal corrosion monitoring and control; aesthetic water quality parameters; customer relations; and system flushing.
Integrated management system (ISO 9000, ISO 14001 and HACCP)	Integrated management systems are becoming a new trend in the water industry where one quality assurance system covers all business management aspects, including general quality management (ISO 9000), protection of the environment (ISO 14001), drinking water safety to the user (HACCP), and worker health and safety. The benefit of implementing one integrated system is that only one audit would be	In Ontario, Canada, a province-wide initiative has resulted in the development of an integrated risk management system based on HACCP, ISO 9001, and ISO 14001 (Kuslikis and White 2004).

(Continued on next page)

Table 3.1 (Continued)
Examples of supporting programs

Program	Description	Example
	required and utility staff will implement only one set of policies and procedures (Deere 2005).	
Data management and control systems	Key documents and records need to be controlled to help support continuous improvement and communication. It is important that the correct versions of operating procedures, checklists and data records are in use. This is simplified today since records and the management of records tends to be based on the use of electronic systems. However, for remote or field workers such systems may have little value and it is important to consider all areas.	South East Water, Victoria, Australia, uses an intranet-based system to support current versions of documents and records. An accountable officer is responsible for maintaining the accuracy of these records and updating them to reflect changed circumstances or to capture improvements. Staff unable to access the intranet system due to working at remote locations, are provided with controlled and registered hard copies of information that can be replaced if required and/or are trained to avoid the need for day-to-day reference to records.
Quality assurance and quality control program for professional services and purchased goods	Quality assurance of vendors and quality control of received goods are an important aspect of reducing the risk of water quality problems. These programs help to minimize contamination of materials and chemicals that come into contact with the drinking water.	Barwon Water, Victoria, Australia, screens suppliers of chemicals and materials and only uses approved suppliers. The suppliers need to describe the quality systems employed to protect product quality and must provide a written agreement that the product is suitable for potable water use. In addition, batches of chemicals are tested from time to time for chemical contaminants.
Legal requirements and best practice	Any legal requirements related to water quality should be complied and risks to breaching compliance are an important consideration in the HACCP plan. Furthermore, best practice and emerging understanding need to be taken into consideration and applied where practicable.	The City of Austin, Texas, maintains a record of the legal requirements related to water quality. The HACCP team identified water quality compliance requirements in its HACCP plan and included the possibility of breaches of these requirements as a risk factor in undertaking its risk assessment.
Operations Programs		
Distribution system operations	The organization needs to commit to developing and using operational procedures in a consistent and reliable manner. The purpose is to use the discipline of documentation to clarify what is expected and to enable peer review.	Sydney Water, New South Wales, Australia, has a quality management system (ISO 9001) in which key operational procedures are documented as SOPs. These SOPs are then cited in the HACCP plan to provide a point of reference for the agreed approach to undertaking important activities. When operational or other staff identifies opportunities to improve operational practices, the SOPs can be updated.
Emergency response program*	If there is a suspicion that the water might have become contaminated, even if that suspicion cannot be readily verified, the utility needs to be able to protect consumers from possibly being harmed by unsafe water. Therefore, a water quality incident response plan needs to be developed to identify how to protect consumers, dispose of suspect water and provide alternatives should an incident occur.	Sydney Water, New South Wales, Australia, has developed a water quality incident response plan in liaison with the health authority. The plan defines who will be responsible for notifying the public, which notification channels will be used, and how the utility will manage the incident to direct any suspect water away from customers and to provide alternative water supplies. Agreements are in place with water haulers and bottled water suppliers in case of such incidents.

(Continued on next page)

Table 3.1
Examples of supporting programs **(Continued)**

Program	Description	Example
Security/ vulnerability assessments	US federal legislation (H.R. 3448, the Public Health, Security, and Bio-terrorism Preparedness and Response Act. 2002), developed in response to recent terrorism events, mandated that water systems serving more than 3,300 people complete security vulnerability assessments and emergency operation plans. In these assessments, utilities evaluated the required level of physical security and how it will be integrated with existing operational activities.	One utility's vulnerability assessment determined one significant threat was the intentional breach of physical security measures at the storage reservoirs. Key existing security measures included chain link perimeter fencing topped with barbed wire, locked security gates, traditional raised ladder access to the reservoirs and standard vents. Recommended security enhancements included improved surveillance, improved ladder access protection, and replacement of the existing vents with a more secure design.
Maintenance Programs		
Valve and hydrant inspection programs	In the event of a water quality incident in the distribution system, the water utility needs accurate information on the location and conditions of all system valves and hydrants. This enables operations staff to quickly isolate the area of the system experiencing a water quality problem and to efficiently remove poor quality water by flushing, if required.	ActewAGL, Canberra, Australia, has a valve-checking program where major divide valves are checked every three years to ensure that the distribution system zoning assumptions are accurate. In addition, records are kept to show when tanks are bypassed and any zoning changes are made.
Finished water storage facility cleaning program	Finished water storage facilities should be cleaned on a regular schedule to minimize water quality deterioration.	The Southern California Water Company established a district-wide reservoir cleaning program in 1995 to control algae and sediment build-up in several reservoirs (Kirmeyer et al. 1999).
Water main flushing program	A water main flushing program helps to keep the system clean and free of sediment, to remove stagnant water, and to remove any untreated or contaminated water that enters the system (Kirmeyer et al. 2000).	The Los Angeles Department of Water and Power implemented a focused flushing program to improve water quality with respect to secondary drinking water standards and to reduce customer complaints (Kirmeyer et al. 2000).
Sanitary procedures (SSOPs†)	Water supply systems need to be kept clean and this involves both scheduled cleaning of mains and water tanks as well as ensuring that when work is done on the system, the activity does not present an additional risk to water quality. The nature of any cleaning agent used needs to be considered as do risks from workers and equipment that may have become contaminated by working on sewer spills.	South East Water, Victoria, Australia, employs contractors to undertake tank and mains cleaning activities. Divers that clean tanks wear contained suits that are sanitized prior to use. When water mains are repaired, the work is done in a pumped-out trench to keep dirt out of the opening and high velocity flushing is used to clear any debris once the repair is made. Water and sewer trucks and pumps are kept separate to reduce cross-contamination risks.
Vermin control	Rats, mice, lizards, frogs, roaches, and birds can all carry bacteria and there have been some examples of waterborne disease outbreaks arising where one or more of these animals was thought to have caused contamination via water tanks. Therefore, minimizing their presence around water assets is important.	Power and Water Corporation, Northern Territory, Australia, keeps reservoirs roofed to keep out birds and other vermin. Gaps in hatches and vents are protected using features such as spikes and mesh to prevent or discourage bird and animal habitation. Branches are kept clear from overhanging tanks to reduce access to those tanks and reduce droppings.

(Continued on next page)

Table 3.1
Examples of supporting programs (Continued)

Program	Description	Example
Training and Certification Programs		
Employee training programs	Utility staff and any contractors employed to work on the system need to be experienced and adequately trained for their specified activities. The training needs to include water quality and public safety, security, employee ethics, sanitary procedures, hydraulics, water sampling procedures, and orientation for new staff.	Water utility employees need to understand the hydraulic flow paths in a distribution system in order to properly protect water quality, and to help protect consumers if water quality problems are suspected.
Operator certification program	The 1996 Amendments to the Safe Drinking Water Act requires that all operators of public water systems be certified.	ActewAGL, Canberra, Australia, has a competency system whereby qualified assessors certify the competency of staff for specific tasks following demonstration of that competency through a practical demonstration. Until certified as competent for a particular task, a staff member is not permitted to work unsupervised on that task.
Community programs		
Community awareness	The community needs to be made aware of how to properly use and handle water. The importance of backflow prevention and of proper management of water extracted from hydrants by water carriers are examples of the types of issues that require community education.	Sydney Water, New South Wales, Australia, manages a backflow prevention program in which the importance of preventing backflow of hazardous substances into the distribution system is communicated through training and awareness campaigns. User-friendly, illustrated education material is produced and distributed via the internet and directly.
Customer relations	If customers notify the utility of poor water quality then, by definition, there could be a problem with the water quality for which the utility is accountable. It's possible the problem arises in the customer's own plumbing system but as a precaution, the underlying cause of water quality complaints needs to be investigated and future recurrences prevented through systematic changes.	South East Water, Victoria, Australia, has a database of all customer complaints and sends out field staff to investigate any complaint not readily resolved by a simple flush of an outside tap or that cannot be resolved over the telephone. The duty officer is notified of clusters of water quality complaints in any one location to enable a full investigation to take place to see what could have caused the problem and to look at making changes to prevent future, similar problems.

* HACCP registrars tend to refer to "Product Recall" procedures but for water utilities shutting off supply completely is often not appropriate due to the vital importance of sanitation and industrial uses. Therefore, water utilities tend instead to adopt a more complex response.
† HACCP registrars often refer to these as 'Sanitation Standard Operating Procedures" (SSOPs) to refer to the SOPs that are in place to document the way that the organization ensures its operations do not introduce hazards.

HACCP registrars unfamiliar with the water industry but familiar with certain sectors of the food, beverage, pharmaceutical, and medical products industries often refer to "Good Manufacturing Practice" (GMP) and "Good Hygiene Practice" (GHP) as part of the Supporting Programs of a HACCP program (WHO 1997). This can cause confusion since GMPs have a specific meaning in an increasing number of industries and relate to certain manufacturing requirements for some types of foods, beverages, drugs, and medical products as required by the FDA in the U.S. Registrars may need educating in the regulatory requirements for the water industry since the term "GMP" is not used in the water sector outside of the packaged (bottled) water products industry and the EPA, not the FDA, is the principal regulatory agency in the U.S. Referring to how the utility meets both the EPA and state requirements for safe production of water acceptable to consumers is the best way to explain how the water industry undertakes something analogous to GMP.

In general, Supporting Programs are formalized and are subject to review and audit as part of the HACCP plan. Very often, SOPs are developed to formally capture Supporting Programs as well as other aspects of the HACCP plan. Each SOP should detail its owner (the position responsible for its maintenance and communication), its objective, the procedures, corrective actions required if the procedures are not followed, and any reporting activities. Examples of specific SOPs for the water distribution system may include:

- Water quality sampling procedures;
- Monitoring procedures during flushing;
- Sanitary procedures to apply during operations such as disinfection before returning facilities to service after maintenance (see also Sanitation Standard Operating Procedures (SSOPs) in Table 3.1);
- Calibration procedures for on-line monitoring devices in the distribution system;
- Specifications for chemicals and raw materials;
- Security procedures for restricting entry of unauthorized personnel;
- Maintenance procedures for operational equipment; and
- Designation of equipment as restricted to use on the water system.

Gold Coast Water, Queensland, Australia has developed a number of SOPs to support their HACCP Plan. SOPs for the distribution system include:

- Distribution water quality analysis and interpretation;
- Investigation of dirty water complaints;
- Reservoir inspections; and
- Rechlorination of reservoirs.

APPLYING HACCP TO WATER DISTRIBUTION SYSTEMS

Each of the twelve steps in developing a HACCP system is explained in detail in the following sections. Specific examples from Project pilot studies and case studies from the water industry are cited where appropriate. The reader is also referred to the utility HACCP Plans in Appendix B and the utility case studies in Chapter 5 for additional examples.

Step 1: Assemble HACCP Team

The HACCP team is responsible for the planning, development, verification, and implementation of the HACCP system. A team that can collectively provide a broad range of expertise, experience, and skill in all relevant areas of the water distribution process needs to be assembled to develop the HACCP plan. Members of the team should therefore come from across the business, including the strategic planning, operational, and design/development sections of the organization. External support from other organizations may be beneficial to provide expertise and experience in the most efficient way to develop HACCP and in specific technical areas. Practicality is as important as technical validity so both operational staff and engineering and scientific staff need to be involved. Individuals with appropriate knowledge of the chemical, physical, and microbiological hazards in water supplies, and the control measures used to manage them, are essential members of the HACCP team.

If the appropriate level of expertise and experience is present on the team there are two benefits. First, it is less likely that important risks will be over-looked. Second, it is less likely that unnecessarily stringent or impractical risk management controls will be implemented. For both reasons, the resource implications of assembling a good team are likely to represent a sound investment.

For consecutive systems, representatives from the bulk supplier should be consulted in the development of the HACCP system. This is imperative, as the condition of the water upon receipt will have bearing upon its management in the distribution system.

If the maintenance and repair of the water distribution system is outsourced, the contracting organization(s) should make commitments to the HACCP process that are equivalent to those made by the client organization. In addition, key persons from the contracted company(ies) should be part of the HACCP team when developing and implementing the HACCP system. This also applies if the sampling and testing of the water for compliance is undertaken by another organization since sound analytical practices are integral to validation and verification of the HACCP plan.

Team members' names, their positions within their organization, their role within the HACCP team and their contact details should be listed in a simple table or spreadsheet (see example in Table 3.2). Formal training in HACCP is often the first step of the HACCP planning process as training imparts a sound common base of understanding among team members.

Table 3.2
Example roles of HACCP team members

Name*	Position in Organization	Role in HACCP	Contact Information (e.g. Email Address, Telephone no.)
	Water Supply & Operations Manager	Project Director†	
	Water Quality Engineer	Project Manager†	
	Field Services Quality Controller	Expert on distribution system work practices	
	Water Distribution Engineer	Expert on distribution systems	
	Maintenance Manager	Expert on maintenance and calibration	
	Trade Waste Manager	Expert on backflow prevention	
	Field Services Supervisor	Expert on service reservoirs	
	Department of Health	Advisor on public health issues	
	Consultant	HACCP plan training and coaching	
	Consultant	Water quality technical advice	

Source: ActewAGL, Canberra, Australia, April 2005

*Government and utility staff details withheld for security reasons.

†Registrars often refer to those with the principal lead or coordination roles as "HACCP Champions."

In larger utilities, to assure the effectiveness of the HACCP system, a core team of more experienced personnel is formed to direct the overall process. This team would include a HACCP Coordinator (Project Manager). Once the core team has been assembled and appropriately trained, working parties would be formed to develop components for their specific section of the water distribution system. Each group would be formed based on its specialist areas and should include representatives for all staffing levels from front line operators through experienced experts.

South East Water, in Victoria, Australia created five working parties in addition to the core team for its HACCP plan which included:

- Treatment systems (booster chlorination and spot dosing);
- Service reservoirs;
- Distribution system;
- Backflow prevention; and
- New extensions.

An executive-level HACCP Champion (Project Director) might be identified to provide high-level support and promotion for the HACCP plan. This can be important in multi-divisional organizations since staff from more than one division will need to work together and will need cross-organizational support. In general, the Chief Executive or General Manager of the whole water utility or water division should be in full support of HACCP implementation before the process begins. One benefit of HACCP for senior management is that HACCP certification can provide a tangible and publicly recognizable endpoint to the process. Organizational leaders can rightly take credit in demonstrating that they are running an operation to a certifiable level of rigor.

Scope of the HACCP Plan

The scope of the HACCP plan should be identified. Some utilities focus on one part of the system such as the treatment facilities or the distribution system, whereas other utilities may wish to prepare a HACCP Plan for the whole system, from source to connection. For example:

- Many utilities, such as Yarra Valley Water and South East Water, Victoria, Australia, identified the endpoint of their HACCP Plan as the endpoint of their legal responsibilities which is often at the customer water meter.
- Other utilities, such as the City of Austin, Texas, and Sydney Water, New South Wales, Australia, developed a HACCP plan that only covered one of their delivery systems, for the purposes of completing a pilot study for this project.
- Power and Water Corporation, Northern Territory, Australia, developed a HACCP plan for one of its systems, source to tap, for a pilot study conducted for this project.

Hints:

- Be open about the resource implications to prevent subsequent resentment if staff find they have committed to more than they anticipated.
- Include, rather than exclude, staff members that are initially skeptical so that they can understand the process and buy into it once they become part of the team.

- Include operational staff and contractors to make sure the plan is practical and sensible.
- Remember that the costs of getting the right team with the right experience are likely to be much less than the costs of struggling with the HACCP plan due to having a weak team.

Step 2: Describe Drinking Water

The drinking water being supplied by a water utility may be referred to as "tap water," "town water," "potable water" or other similar terms. However, for the development of a HACCP plan, more detail needs to be provided as a basis for the risk assessment in Step 6. Specifically, information is needed on the regulatory requirements and utility goals that affect the quality of the finished drinking water supplied to customers, and the so-called "inputs", such as treatment chemicals.

One of the first required statements refers to the specific quality of water being supplied. This is important because some specialized industrial or medical uses of water may require further treatment beyond that provided by the water utility. The starting point is the regulatory and contractual requirements that specify the minimum standards that the water must meet. These should be explicitly referenced and may include:

- EPA federal regulatory requirements;
- State regulatory requirements;
- Operating licenses;
- Customer charters or contracts;
- Water supply agreements; and
- Health guidelines.

A general description of the source of the water, the type of treatment used, the storage conditions, and the method of distribution needs to be specified. The raw materials used in treatment and operations, such as disinfectants, need to be specified since they effectively become part of, or could contaminate, the water as supplied. Material Safety Data Sheets can often provide evidence that the concentrations as supplied are safe for potable use.

Water utilities often distribute more than one type of water such as recycled water, potable water and raw water. All of these types of drinking water can be covered under a HACCP Plan, as illustrated in the examples in Tables 3.3 to 3.5.

Table 3.3
Example descriptions of drinking water

Product Group: Drinking water (also known as "potable," "tap," "town" water)

Specifications: ActewAGL produces drinking water with the specific water quality specifications being set out in Contracts with ACTEW and cited documents and codes such as the Drinking Water Quality Code of Practice 2000 (ACT Health) and the Drinking Water Utility Licence 2005 (ACT Health).

Process: Water is harvested from the Cotter River and Googong River surface water catchments. Water is abstracted directly from one location on each of the Bendora and Lower Cotter open reservoirs on the Cotter River and from the Googong open Reservoir on the Googong River. Water is treated via filtration and disinfection and distributed to customer water meters via a closed piped reticulation which includes enclosed balancing storages and pump stations.

Scope: ActewAGL becomes directly responsible for the water once it is present in Bendora, Lower Cotter or Googong Reservoirs. ActewAGL is no longer directly responsible for the water once it has passed through customer water meters.

Inputs*: Treatment chemicals are added at the treatment plants and at booster disinfection points in the distribution system (description omitted for brevity). The raw material is surface water harvested as described under Process, above. For booster chlorination, 12% sodium hypochlorite liquid is automatically dosed into three distal reservoirs.

Product Group: Raw water (also known as "bulk," "river," "untreated" water)

Specifications: ActewAGL produces raw water with no nominated specifications.

Process: Raw Water is harvested from the Cotter River, abstracted from the open Bendora reservoir and distributed untreated to customers via a connection to the closed piped bulk transfer system.

Scope: ActewAGL becomes directly responsible for the water once it is present in the Bendora and Lower Cotter Reservoirs. ActewAGL is no longer directly responsible for the water once it has passed through customer water meters.

Inputs*: 65% calcium hypochlorite granules are manually dosed into reservoirs. No other chemicals are added to the raw water. The raw material is surface water harvested as described under Process, above.

Source: ActewAGL, Canberra, Australia, April 2005. Reprinted with permission.
*Often referred to as "ingredients" or "preservatives" by HACCP registrars not familiar with HACCP application to water.

Table 3.4
Example specifications for potable water

Analyte	Specification					
*E.coli**	<1 organisms/100mL in 98% of samples over a 12-month period. (All distribution zones)					
Free chlorine residual	95% UCL† for samples over a 12-month period ≤ 5 mg/L					
Total chlorine residual	95% UCL† for samples over a 12-month period ≤ 5 mg/L					
Ammonia	95% UCL‡ for samples over a 12-month period ≤ 0.5 mg/L					
Aluminium (soluble)	95% UCL‡ for samples over a 12-month period ≤ 0.2 mg/L					
Color (true)	95% UCL‡ for samples over a 12-month period ≤ 15 HU§					
Copper	95% UCL‡ for samples over a 12-month period ≤ 1 mg/L					
Fluoride	95% UCL† for samples over a 12-month period ≤ 1.5 mg/L					
Hardness	200 mg/L 95% UCL‡ for samples over a 12-month period ≤ 200mg/L					
Iron (mg/L)	95% UCL‡ for samples over a 12-month period ≤ 0.3 mg/L					
Manganese	95% UCL‡ for samples over a 12-month period ≤ 0.1 mg/L					
pH	6.5 – 9.2 within 95% confidence interval for samples over a 12-month period and no individual sample greater than 11					
Trihalomethanes (THMs)	95% UCL† for samples over a 12-month period ≤ 0.25 mg/L					
Turbidity	95% UCL‡ for samples over a 12-month period ≤ 5 NTU					
Electrical conductivity	95% UCL‡ for samples over a 12-month period ≤ 1000 μS cm^{-1}					
Flow rate	Diameter of property service pipe (mm)	20	25	32	40	50
	Minimum flow rate (L/min)	20	35	60	90	160
Interruption to supply	Not to exceed 5 hours in 95% of instances over a 12-month period					

Source: Yarra Valley Water 2005. Reprinted with permission.
*Organism = a colony forming unit (cfu)
†UCL = Upper confidence limit of the 95th percentile of annual samples
‡UCL = Upper confidence limit of the mean of annual samples
§HU = Hazen units

Table 3.5
Example specifications for non potable water

Parameter	Specification
Intended use	Non-potable water not intended for human consumption†, ‡ Reference: NHMRC/NRMMC 2004.

Source: Yarra Valley Water 2005. Reprinted with permission.
*UCL = Upper confidence limit of the mean of annual samples
† Untreated water in bulk supply lines controlled by Melbourne Water. Customers in these areas have been informed of the non-potable quality of their supply and instructed of its intended use.
‡ The Department of Human Services (DHS) issued a Drinking Water Regulation Guidance Note 1 in April 2005 under Section 6 of the Safe Drinking Water Act. This note states that "Supply-by-agreement" arrangements cannot be declared as regulated water if there is a clear intent that the water is to be solely or partly used as drinking water. We will develop a policy for this customer group to meet the requirements in this Guidance Note in consultation with the customers and the DHS. Our current controls ensure that the water could not be mistaken as drinking water.

Hints:

- Include all water quality parameters, i.e. chemical, physical, microbiological, and radiological.
- Focus on drinking water supplied to the majority of consumers.
- Consider specific customer groups that exist such as customers supplied via private extensions who may receive non-potable water.
- If the final drinking water specification is set by a regulating body, consider referencing this in the HACCP plan rather than including it directly, this ensures the HACCP plan does not become outdated if the regulatory requirements change.
- Identify any chemicals that are added to the water.

Step 3: Identify Intended Use

The statement given here needs to clarify who are the intended users of the water and how they will use it. In general, the most sensitive water use recognized by water utilities is that of potable water (hence the logic of applying HACCP to water). In such a case, the intended use statement needs to clearly state that the water is intended for domestic uses without further treatment by the user, including washing, drinking and food and beverage preparation, by the majority of the population. In some cases, other types of drinking water are supplied that may have different intended uses.

Some specialized industrial or medical uses of water are unlikely to be satisfactorily serviced by conventional municipal water supplies and this needs to be clearly stated in the HACCP plan. It is important to highlight any exceptions to the general case. For example, dialysis patients or severely immunocompromised persons may need to undertake additional treatment before using the water. There may be community groups that are singled out for special considerations. There may be some private extensions to the water supply which may tap off prior to completion of treatment or after excessive in-pipe residence times.

By stating that the water is intended to be consumed by the majority of the population, it excludes the necessity to cater to minority groups with special needs or industries with specific requirements. These users are likely to be advised by medical professionals (medically identified groups) or process engineers (industrial customers) if their water received at their tap may require further treatment such as boiling or filtration.

In some cases the water utility may think it prudent to make some explicit statements as part of its community education Supporting Program about the intended use and users of its drinking water. For example, aquarium owners often do not realize that potable water is not always safe for fish due to the presence of chlorine. Using internet sites, rates notices and pamphlets, it is possible to notify customers of the quality of the water supplied to them. For example, Sydney Water, New South Wales, Australia, sends out quarterly leaflets detailing water quality compliance results and providing a general statement regarding the water quality supplied.

Because potable water is such a familiar and common place item, extensive statements are not required. Nonetheless, the amount of detail provided in utility HACCP plans varies. Examples of intended use statements are provided in Table 3.6 and Table 3.7.

Table 3.6
Example of a long-form "intended use" statement

Drinking water.
Intended Uses: Drinking water is intended for general consumption as a beverage, for washing of bodies and clothes and the preparation of beverages and foodstuffs.
Intended Users: For use by the general population of Canberra, the Australian Capital Territory, and Queanbeyan.
Unintended uses and users: There are some industrial, agricultural and commercial uses for which ordinary drinking water is unsuitable and such customers are not necessarily the intended users. There are people that are advised to provide additional point-of-use treatment before drinking water based on specific medical advice and such patients are not necessarily the intended users. Drinking water is not intended for use by those other than customers holding water services accounts and known to be connected to the distribution systems or holding permits to draw potable water from hydrants.
Raw water.
Intended Uses: No specific purpose.
Intended Users: Customers are at Pierces Creek, Uriarra Forestry Settlements and at four rural properties upstream of Stromlo WTP.
Unintended uses and users: Raw water is not intended for consumption without first applying treatment. Raw water is not intended for customers other than those formally recognized by supply agreements.

Source: ActewAGL 2005. Reprinted with permission.

Table 3.7
Example of a short-form "intended use" statement

Intended Use	Potable water intended for general human consumption.

Source: Yarra Valley Water 2005. Reprinted with permission.

Hints:

- Be specific when stating intended uses of the water for each customer group.
- Cater to the majority of consumers.
- Consider specific customer groups such as customers supplied via private extensions.
- Provide instructions to accommodate unique customers.
- Consider alternative methods to notify sensitive users such as an internet site, a written notice (bill stuffer) and as part of a Community Relations Program.

Step 4: Construct Flow Diagram

The HACCP team develops a flow diagram that depicts processes and operations that take place throughout the water distribution system. The diagram should illustrate what happens to the water from the time it is received (either at the catchment or from the bulk supplier) until it reaches the customers' taps. It should include any points where the water is changed or controlled.

Key features of the distribution system that should be included in the flow diagram are pump stations, chemical feed points, storage tanks, and control valves. The flow diagram is an important basis for the hazard analysis in Step 6. The flow diagram should contain adequate detail to identify potential entry points for hazards and to trace any detected contamination through to subsequent process steps.

The level of detail required for the flow diagram is often a matter of opinion. Examples of a more complex and simpler flow diagram are given in Figure 3.1 and Figure 3.2, respectively. A key point is that processes and points that are common across many parts of the distribution system can be identified as just one symbol on one diagram if the risks and controls associated with those processes or points are common. Another key point is that hydraulic system diagrams and network maps and diagrams are likely to already exist. The purpose of the flow diagram is not to duplicate these, but to provide a higher-level process flow diagram at the conceptual level.

Once the critical control point and monitoring steps (Steps 7 and 9) are complete, the flow diagram will be used to indicate CCPs and monitoring points (as in Figure 3.2).

The symbols used in the flow diagram should be straightforward and illustrate key processes:

- Storage ("open" or "closed" such as tanks, reservoirs, basins);
- Transport (water is moved from one place to another either by gravity or pumping);
- Inspection points (monitoring occurs often resulting in a decision); and
- Operation (an intentional change occurs such as disinfection).

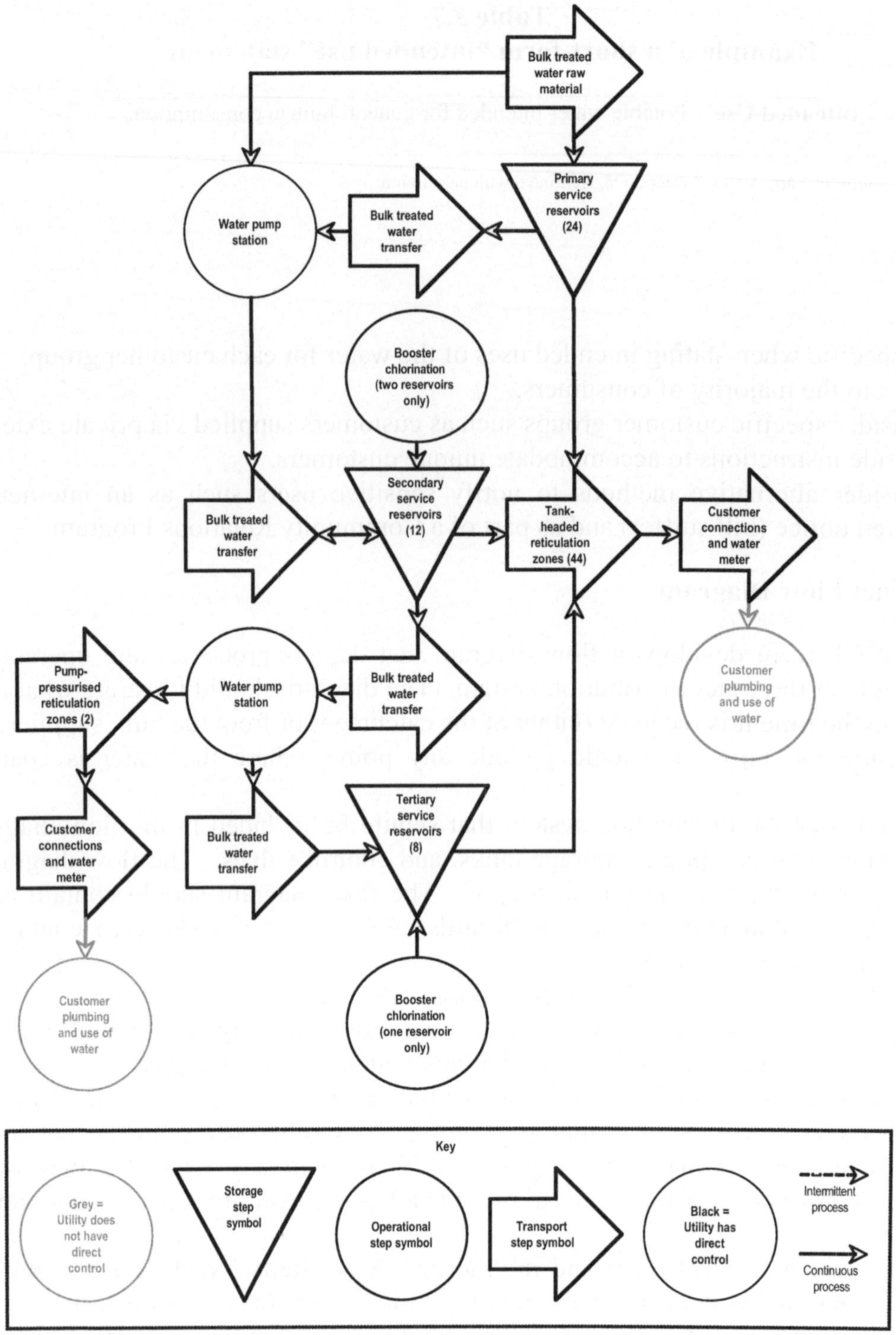

Source: ActewAGL 2005. Reprinted with permission.

Figure 3.1 Example of more complex system flow diagram

	Description	Symbol	Details
CCP	Treated water receival from wholesaler	○	Downstream of treatment
	Transfer	⇨	> 750 mm diameter
CCP	Storage	▽	Closed storage
	Distribution	⇨	300 mm to 750 mm diameter
CCP	Reticulation	⇨	< 300 mm
	Customer Interface	⇨	

Source: Yarra Valley Water 2005. Reprinted with permission.

Figure 3.2 Example of simpler system flow diagram

Hints:

- Think in terms of "water quality processes" rather than "water quantity processes."
- Produce more specific flow diagrams as subordinate flow diagrams for each individual section to provide greater detail if necessary.
- Update flow diagrams when facilities are added or reconfigured.
- Keep the flow diagram as simple as possible to avoid confusion.

Step 5: Confirm Flow Diagram

In Step 5, the accuracy of the flow diagram is checked by reviewing that all operations in the water distribution system are being considered and evaluated. This step helps to reduce re-work at a later point if gaps are subsequently found. One of the most time-efficient means of validating the flow diagram is to step through the diagram with experienced system operators and water quality engineers or scientists.

If there is uncertainty about aspects of the flow diagram, physical inspection can be undertaken. In terms of the fine details, it is not necessary to verify every valve and service pipe location but it is worth referring to the maps and plans wherein such information is kept. A program of valve checking is recommended as part of a HACCP program to ensure that assumptions about zoning are valid and that water is not, in fact, transported in different directions from what is documented on the detailed system diagrams.

Once the preliminary steps (Steps 1-5) have been completed, the core HACCP steps can be applied (Steps 6-12).

Hints:

- Sign and date the flow diagram to demonstrate its veracity.
- Update the flow diagram(s) when new facilities are added or existing facilities are reconfigured.

Step 6: Conduct a Hazard Analysis

Step 6 (Conduct Hazard Analysis) includes three elements: hazard identification, hazard assessment, and identification of control measures.

Hazard Identification

Once a HACCP team is formed and the preliminary steps completed (e.g. constructing and validating the process flow diagram), "…the HACCP team should list all of the hazards that may be reasonably expected to occur at each step from …distribution until the point of consumption" (Codex Alimentarius Commission 1993). To perform this step, the HACCP Team relies on their experience, system knowledge, historical data, and current technical knowledge. The flow diagram provides the frame of reference for this task with each symbol on the flow diagram representing a process step. The following procedure is undertaken for each process step:

- The process step is listed;
- The causal events (hazardous events) by which hazards can arise are identified; and
- The types of hazards that can arise are listed.

In identifying potential distribution system water quality hazards, the HACCP Team should consider whether the elimination or reduction of the hazard to an acceptable level is essential for water to be considered potable.

Four types of general hazard categories are used: physical, chemical, microbiological, and radiological. Examples of each category are specified in Table 3.8. In practice, the radiological category is not always considered.

Table 3.8
Examples of hazards to the water distribution system

Hazard	Examples
Physical	Dislodged pipe tubercles, dirt, tree roots, fragments of pigs and swabs
Chemical	Gasoline, oil, industrial discharges, cleaning fluids, threat agents*
Microbial	Bacteria including coliforms and *E. coli*, threat agents†
Radiological	Uranium, radium, threat agents

* Chemical threat agents for drinking water systems include chemical warfare agents (e.g. hydrogen cyanide, sarin, and sulfur mustard) and industrial chemical poisons (e.g. cyanides, arsenic, fluoride, cadmium and mercury) (Mays 2004).
† Biological threat agents for drinking water systems are potentially resistant to disinfection, can be produced and disseminated in large quantities, and are stable for relatively long periods in water (Mays 2004). These biological threat agents include certain waterborne pathogens (e.g. *Clostridium perfringens*, plague) and biotoxins (e.g. botulinum, aflatoxin, ricin).

Examples of potential hazards to a water distribution system are summarized as follows:

- Contamination due to treatment failure;
- Fecal contamination of storage tanks and standpipes;
- Backflow event causing contamination;
- Contamination caused by negative or low pressure transient resulting in intrusion of untreated water through submerged air valve, faulty seal or point of leakage;
- Contamination caused by fecal material or by foreign inanimate objects (e.g. dirt, mud, timber, plastic) during repair, alterations, or connection to existing mains;
- Contamination by hazardous substances as a result of use of products, materials or coatings that are not approved for contact with potable water; and
- Contaminants introduced by permeation, pipe degradation or corrosion.

The system supplying water to Zurich, Switzerland considers four categories of potential risks (Bosshart 2003):

- Personnel that are not properly trained or experienced;
- Materials in contact with drinking water;
- Machines in contact with drinking water; and
- Faulty methods or procedures related to distribution system management.

The Swiss regulatory guideline (W1002) entitled "Recommendations for a Simple Quality Assurance System for Water Supplies includes checklists of possible hazards (Swiss Gas and Water Industry Association 2003). Checklist items that pertain to distribution system reservoirs include the following hazards:

- Natural risks (e.g. trees, landslides, animals, vermin);
- Unsecured access (e.g. doors, windows, ventilation);
- Poorly separated water chambers;

- Poor building condition (e.g. concrete, coating, piping);
- Poor water circulation (e.g. fire water reserve);
- Poor ventilation (e.g. water chamber, dehumidification plant);
- Pollution (e.g. wastewater, soiled clothing);
- Unsuitable process control (e.g. no displacement of inlet); and
- Materials (e.g. chemicals, cleaning agents).

Checklist items that pertain to piping networks include the following hazards (Swiss Gas and Water Industry Association 2003):

- Natural risks (e.g. line breaks);
- Unsecured access (e.g. hydrants, gate valves);
- Unfavorable pressure conditions (e.g. suction of external water);
- Impaired functionality (e.g. motors, controllers);
- Blocked access (e.g. shut-off valves);
- Missing or incorrect flushing (e.g. mains, hydrants, fountains);
- Improper repair work and connections;
- Unsecured customer installations (e.g. immersions, pressure pumps); and
- Consumer behavior (e.g. seasonal operation).

Hazard Assessment

Most U.S. water utilities are familiar with hazard assessments from their experiences with completing federally mandated vulnerability assessments and developing cross-connection control programs. For example, in developing a cross-connection control program, the utility identifies and rates possible cross-connections as low, medium or high hazards, and then installs (or requires the customer to install) backflow prevention devices for high hazard locations.

In assessing the risk of a hazard, the following criteria should be evaluated:

- Frequency of occurrence; and
- Severity of consequences (possible in terms of size of the population affected, the type of impact and the duration of the hazardous effect).

In undertaking the hazard analysis, it is possible to adopt one of three alternative approaches:

- A simple qualitative approach;
- A simple semi-quantitative scoring approach; or
- A complex semi-quantitative scoring approach.

The utility's HACCP team should agree upon one risk ranking procedure. As suggested by NHMRC/NRMMC (2004), agreeing on the approach before hand and initially testing it with a small group may help assure that an appropriate risk assessment approach is selected.

In the simple qualitative approach, the HACCP team assigns one of the following risk ratings to each hazard based on the judgment of the group: "significant," "insignificant," or "uncertain" if further analysis was required. This simple qualitative approach was adopted by Gold Coast Water, Queensland, Australia.

The simple semi-quantitative scoring approach, illustrated in Table 3.9, involves assigning a score to each hazard for two factors, the hazard's likelihood of occurring, and the severity of consequences that could result from an occurrence (Gray and Morain 2000; Deere et al. 2001). The scoring is based on a 1 to 5 scale. For the likelihood of occurrence factor, a score of 1 is given for a rare occurrence, while a score of 5 indicates a hazard that is almost certain to occur. Each hazard is also scored on the possible severity of consequences that could result, with a score of 1 for insignificant consequences and a score of 5 for catastrophic consequences. These two scores are multiplied to calculate the risk factor rating. This approach was adopted by the participating utilities in this Project, as well as by Brisbane Water, in Queensland, Australia and South East Water and Yarra Valley Water, in Victoria Australia. As an example, the risk of a pathogen (the hazard) contaminating the water and reaching hazardous levels as a result of a water main that was repaired and returned to service (the causal hazardous event) at the distribution process step (not being properly flushed) might be considered to be unlikely (likelihood factor of 2) but have a severity of consequence of 5 (public health risk). The product of these scores gives a risk factor rating of 10. Two additional examples are shown in Tables 3.10 and 3.11.

Table 3.9
Example of a simple risk-scoring matrix

		Severity of Consequences				
Risk Factor Matrix:		Insignificant *No impact / not detectable* *Rating: 1*	Minor *Compliance Impact* *Rating: 2*	Moderate *Aesthetic Impact* *Rating: 3*	Major *Regulatory Impact* *Rating: 4*	Catastrophic *Public Health Impact* *Rating: 5*
Likelihood	Almost Certain *Once a day*	5	10	15	20	25
	Likely *Once a week* *Rating: 4*	4	8	12	16	20
	Moderate *Once a month* *Rating: 3*	3	6	9	12	15
	Unlikely *Once a year* *Rating: 2*	2	4	6	8	10
	Rare *Once every 5 years* *Rating: 1*	1	2	3	4	5

Source: Adapted from AS/NZS 4360:1999 (superceded); AS/NZS 4360:2004; and HB 436-2004 ("the Works") with permission from SAI Global Ltd. These standards can be purchased at **http://www.sai-global.com**

Using this method of evaluating risks, a risk is often deemed to be significant if the rating is 6 or greater. The cut-off of 6 may be selected to avoid focusing on risks that will be inconsequential due to rarity or low severity. Depending on available resources, the HACCP team decides how many hazards can be addressed by the HACCP Plan (e.g. all hazards with a risk score >10 might be addressed) and how they will be prioritized.

Table 3.10
Example of a risk assessment for service reservoirs

Process step: Closed service reservoirs

Risk (hazard and cause)	**Likelihood**	**Consequence**	**Risk level** (max)	**Preventive Measures** (that reduce risk to acceptable level)	**Comments** (basis for assessment and supporting programs)
Birds or rodents intrude through vermin-proofing then defecate in water and release pathogens or microbial indicators exceeding guidelines	3	8	Very High	• Vermin proofing of storages • Residual disinfection maintained in storages	• Covered under "pest control" supporting program • The same applies to any open storage system and risks are higher closer to sources of human pathogens, such as refuse tips
Unsanitary maintenance and repair leading to introduction of hazards exceeding guidelines	1	3	Low	• Adherence to sanitary maintenance and repair procedures	• Repairs and maintenance are only annual or less on service reservoirs
Deliberate contamination by tipping of hazardous substance through hatch to levels exceeding guidelines	2	8	Very High	• Security inspections • Locked ladder shroud • Locked hatch • Fencing	• The logistical difficulties, dilution factors, and short duration make such an act both difficult and of limited maximum consequence but there have been examples of this taking place which resulted in major public confidence impacts
Loss of pressure below minimum service level leading to ingress of contaminants through back siphonage to levels exceeding guidelines	2	3	Moderate	• Pumps re-fill tank at low level set point through automated feedback loop	• Dilution factors, short duration and subsurface location of assets makes such an event of limited maximum consequence
Unsafe or inappropriate materials used in making repairs or adding fittings leading to introduction of hazards exceeding guidelines.	1	3	Low	• Use of approved materials • Adoption of water industry codes for materials and construction as per prerequisites	• Covered under "inputs quality assurance" and "best practice" supporting programs • Dilution factors make such an event of limited maximum consequence • Materials for service reservoirs are likely to have been produced specifically for that purpose
Water becomes stagnant leading to taste and odour compounds and microbial indicator levels exceeding guidelines	3	3	High	• Pumps refill tank at low-level set point through automated feedback loop to encourage turnover • Drainage and isolation of redundant tanks during winter	• Not a health issue but a compliance issue

Source: ActewAGL 2005. Reprinted with permission.

Table 3.11

Example of a risk assessment for distribution systems

Process step: Closed treated water distribution

Risk *(hazard and cause)*	**Likelihood**	**Consequence**	**Risk level** *(max)*	**Preventive Measures** *(that reduce risk to acceptable level)*	**Comments** *(basis for assessment and supporting programs)*
Sloughing of slimes or resuspension of sediment during high flow periods, following shut down and recharge, flow reversals or pipeline disinfection and cleaning leading to iron, manganese and turbidity exceeding guidelines	4	2	High	• Controlled valve operation • Management of flow velocity and direction • Systematic zoning and network operation • Scheduled mains cleaning to remove sediment	• Variable and often low velocity in pipelines and small size of pipelines makes this event likely • This is not a health risk if it does occur
Deliberate contamination by injection of hazardous substance into pipeline to levels exceeding guidelines	1	6	Moderate	• Positive pressure maintained in pipeline • Buried pipes and fittings	• The logistical difficulties make such an act difficult and the short duration and small population affected reduces the maximum risk
Accidental contamination through pressurised cross-connection or service connection or backflow during pressure drop leading to hazardous substance flowing into pipeline to levels exceeding guidelines	2	6	Very High	• Positive pressure maintained in pipeline • Backflow prevention for high hazard connections • Separate sewer and water trenches • Coloured water, sewer and recycled water pipes • Plumbing code requirements and regulations	• Covered under "best practice" and "hygiene and sanitation" supporting programs • The positive pressure makes event unlikely • Short duration and small population affected • Advise plumbing and home handyman sectors
Unsanitary pipeline maintenance and repair leading to introduction of hazards exceeding guidelines	4	6	Very High	• Adherence to sanitary maintenance and repair procedures including thorough flushing of small lines after repair and flushing and disinfection of large lines	• Covered in "hygiene and sanitation" supporting • Repairs are common on smaller. • Short duration and small population affected reduces maximum risk but it can be a health risk
Unsafe or inappropriate materials used in making repairs or adding fittings leading to introduction of hazards exceeding guidelines	1	3	Low	• Use of approved materials • Adoption of water industry codes for materials and construction	• Covered under "inputs quality assurance" and "best practice" supporting programs • Materials used in water supplies are especially produced for that purpose

(Continued on next page)

Table 3.11

Example of a risk assessment for distribution systems (Continued

Process step: Closed treated water distribution

Risk *(hazard and cause)*	**Likelihood**	**Consequence**	**Risk level** *(max)*	**Preventive Measures** *(that reduce risk to acceptable level)*	**Comments** *(basis for assessment and supporting programs)*
Water becomes stagnant leading to taste and odour compounds and microbial indicator levels exceeding guidelines	4	2	High	• Design of system to reduce dead ends and large diameter pipes where not required • Flow management through zoning and operation • Scheduled flushing to remove stagnant water	• Not a health issue but a compliance issue • Small population affected reduces maximum risk
Corrosion in court bowls and service lines leading to iron or copper exceeding guidelines	4	2	High	• Scheduled flushing to remove stagnant water • Stabilization and pH control of water • Maintenance of residual disinfectant • Scheduled replacement of lines	• Covered under "inputs quality assurance" and "best practice" supporting programs • Of only very minor health significance, more an aesthetic issue
pH above guidelines in cement-lined pipes	3	1	Low	• Use of lined fittings for new systems	• Not a health issue
Loss of residual leading to microbial growth	3	1	Low	• Maintenance of residual disinfectant	• Can affect total coliform compliance

Source: ActewAGL 2005. Reprinted with permission.

The City of Austin, Texas identified a number of potential hazards for their pilot study area:

- Backflow through an unprotected cross connection (risk score 25);
- Contamination at new construction sites (risk score 20);
- Backflow from failing septic systems into distribution main (risk score 9);
- Contamination due to water main break or repair (risk score 12);
- Pathogen intrusion to distribution main due to leaky sewage main (risk score 10);
- Intentional contamination of distribution system by vandals (risk score 9);
- Contamination due to main break outside pressure zone (risk score 8);
- Contamination due to pressure transient (risk score 4);
- Presence of bacteria due to low disinfectant residual (risk score 4); and
- Presence of bacteria due to zero disinfectant residual (risk score 4).

The HACCP team decided to base the pilot study on two particular hazard events that scored high marks in the hazard analysis – backflow from unprotected cross-connections and hazards at new construction sites that can potentially degrade the finished water quality. The concern for possible contamination at new construction sites focused on contractors' unauthorized operation of system valves and the effectiveness of disinfection of new mains prior to being placed into service.

A more complex semi-quantitative matrix used by Sydney Water and described in Deere et al. (2001) is illustrated in Figure 3.3.

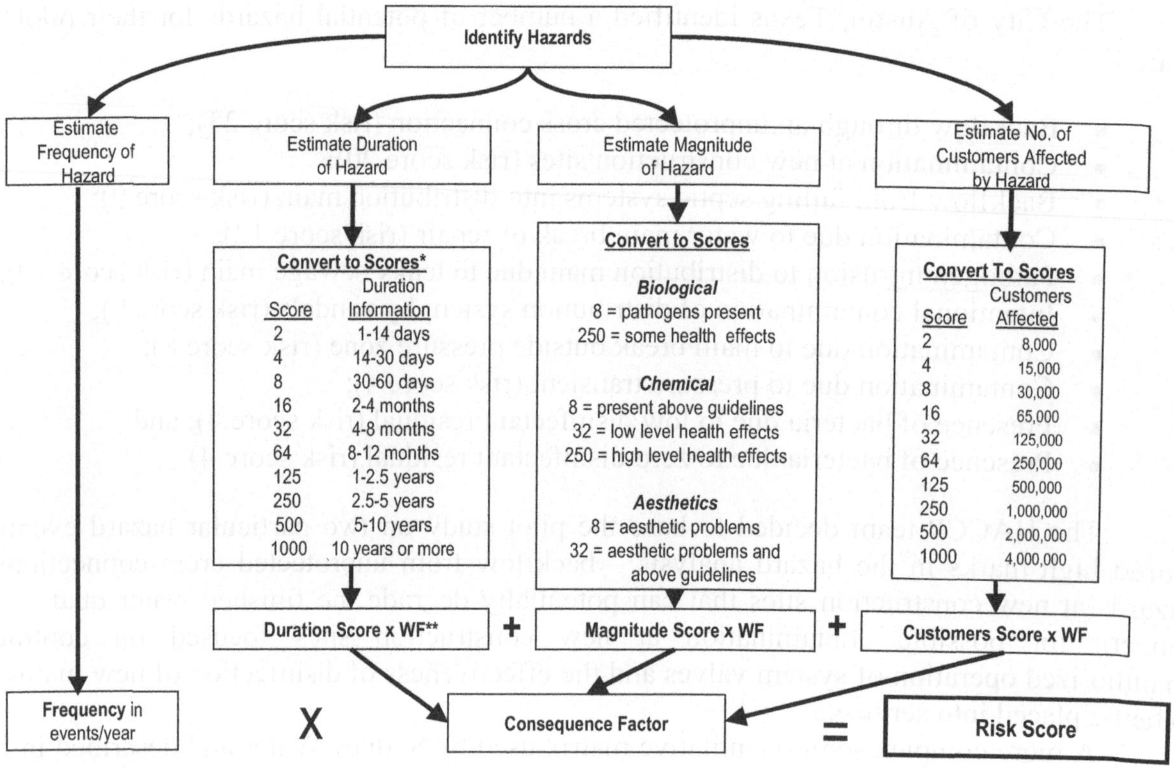

Source: Deere et al. 2001. Reprinted with permission from WHO, **www.who.int**
WF = weighting factor, default value = 0.33

Figure 3.3 Example of a complex risk ranking approach

There are less structured means of prioritizing risks that do not include the use of a risk matrix. If individuals serving on the HACCP team have the skills needed to judge the severity and likelihood of hazards and hazardous events, then risks can be prioritized in a less structured format.

Identification of Control Measures

Once hazards have been identified and ranked according to their likelihood of occurrence and severity of consequences, the HACCP Team focuses on how to eliminate or reduce the seriousness of the highest ranked hazards. This is accomplished with HACCP steps 6 through 10.

In HACCP step 6, the HACCP team identifies existing control measures that are being used within the distribution system, and any additional control measures that systems may consider for future operations. Control measures or preventive measures are defined as all actions taken to prevent or reduce contamination in a drinking water supply. Typical control measures for distribution system hazards include maintenance of positive pressure at all times; use of sanitary procedures during construction; and adequate construction of all storage and distribution facilities. Table 3.12 lists control measures for several hazards identified by South Berwick Water District.

Table 3.12
Example hazard analysis and control measures

Hazard Event	Severity of Consequences (score using 1 to 5 scale with 5 being most severe)	Likelihood of Occurrence (score using 1 to 5 scale with 5 being most likely)	Risk Factor = Likelihood x Severity	Existing Control Measures	Additional Control Measures Recommended by HACCP Team
Backflow through an unprotected cross-connection	4	5	20	• Utility installs double check valves on residential services • Commercial customers install backflow prevention device as required • Utility maintains good records of backflow prevention devices • System static pressure >40 psi (276 kPa) system-wide	• Utility needs to enforce testing of commercial backflow prevention devices • Public education • The feasibility of testing backflow prevention devices at multi-family units should be further evaluated
Contamination via Storage Facility Vents	5	2	10	• Storage tank site fenced and well-maintained • Gravel road to site is gated • Site inspection three times weekly • Insect screening on vents kept in good repair	• On-site security camera • SCADA intrusion alarm
Main Break	5	2	10	• Disinfection and flushing of all main breaks before placing back into service	• Review inspection procedures • Review water quality testing procedures

It is likely that during the risk assessment process, the preventive measures identified will be considered inadequate for some of the risks. In such situations, additional control measures will be identified as being desirable and the HACCP team should record these for further evaluation.

Hints:

- Remember that the risk being assessed is the risk of a hazardous event occurring at that process step and leading to a specific hazard reaching hazardous levels.
- One form of hazardous event is the failure of a control measure. Therefore, assess risks for both the normal situation of control measure operation and the hazardous situation of control measure failure.
- Undertake a reality check at the end of the process to be sure that an appropriate risk rating is assigned.
- Document the reasons for the risk ratings and validate them where there is some uncertainty.
- Remember that one hazardous event such as a burst water main can introduce a multitude of hazards.
- Remember that more than one control measure may be required to control a potential hazard.
- Remember that more than one potential hazard may be controlled by a single control measure.
- Ensure that personnel who will control/prevent the hazard know and understand their requirements.
- Ensure adequate downstream control measures exist if a potential risk is not associated with immediate water quality and safety.

Step 7: Determine the CCPs

Distinct from control measures, a CCP is a step in the water system where control measures are essential to maintain the safety of the drinking water. The CCP can be:

- A process step as identified on the flow diagram;
- A specific operation or point in the system; or
- A procedure.

Therefore, in distribution systems, these critical barriers are defined as CCPs by stating that the CCP is the "Process Step" on the flow diagram, not necessarily a single "Point" in time and space. For example, the "service connection" process step can be a CCP since, at that process step, backflow prevention can be critical. This approach was not preferred by WHO and New Zealand Ministry of Health who have removed the use of the term CCP. However, the Australians have retained it and have recognized the validity of the term and the ease with which experienced HACCP practitioners can apply it through conceptualizing the process via the flow diagram.

All significant hazards in the process should be controlled at a CCP or through a reliable Supporting Program to ensure they do not pose a significant water quality and/or safety risk.

Some water quality impacts that are not health-based but can cause aesthetic impacts, such as taste and odor or manganese slimes, can be controlled at points that could be termed quality control points (QCPs). QCP is not a Codex definition but has been adopted by some HACCP-certified water utilities to ensure appropriate focus on aesthetic water quality as well as safety (e.g. South East Water and Yarra Valley Water, Victoria, Australia). Another term, non-critical control point (NCCP), is not a Codex definition, but has been adopted by Power and Water Corporation in the Northern Territory, Australia to ensure appropriate focus on non-safety water quality issues in general as well as safety.

In practice, making such distinctions between CCPs for quality, safety, and compliance is not essential to HACCP and to do so would be a decision made by the HACCP team. For the purposes of this guidance, CCPs will be referred to with respect to control points for quality, compliance, or safety.

One possible tool for determining whether a particular step in the process is a CCP is the decision tree depicted in Figure 3.4. If a particular step is critical for maintaining safe and potable water then it should be regarded as a CCP. The reasons why this step has been determined to be a CCP should be documented in the HACCP plan.

In practice, utilities differ with respect to whether particular process steps represent CCPs. This partly reflects personal preferences, differences in risk profiles and differences in the way that the HACCP plans are structured. Utility examples are provided in Table 3.13.

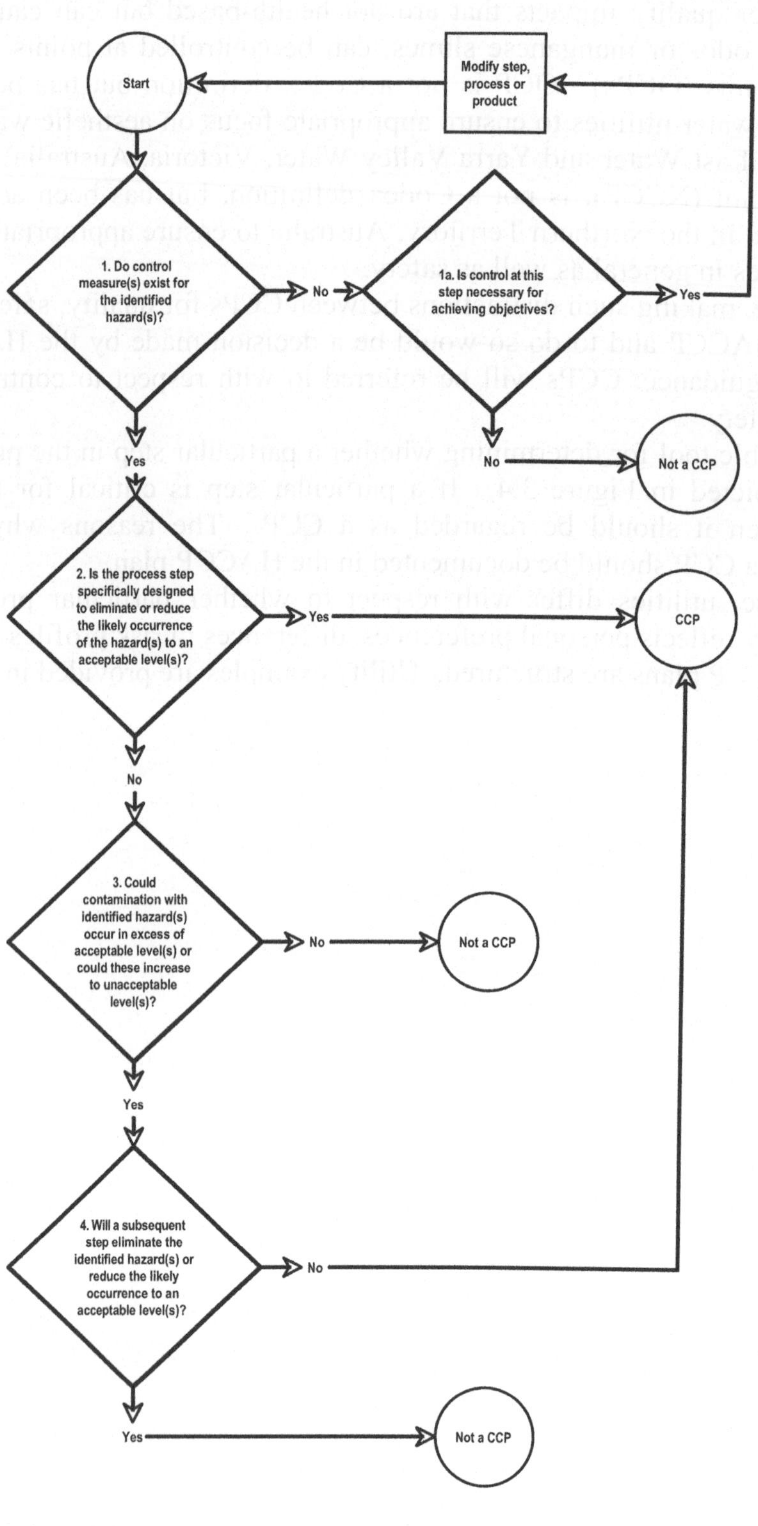

Source: WHO 1997. Reprinted with permission.

Figure 3.4 Decision tree for identifying process steps that are CCPs

Table 3.13
Examples of CCPs in utility HACCP plans

Process, operation or procedure	Example of a utility that has identified this CCP
Interface with bulk supplier	South East Water, Victoria, Australia
Closed reservoir storage	ActewAGL, Canberra, Australia
Chlorination on outlet of open reservoir	Barwon Water, Victoria, Australia
Chlorination on inlet of closed reservoir	Sydney Water, New South Wales, Australia
Maintenance of positive pressure in system	Power and Water Corporation, Northern Territory, Australia
Customer interface	Yarra Valley Water, Victoria Australia
Backflow prevention system	South East Water, Victoria, Australia
New construction processes	City of Austin, Texas

Hints:

- Identify CCPs only for those risks that are significant.
- Record the basis for designating steps as CCPs in the HACCP plan.
- If not using the Codex Decision Tree (Figure 3.4), document the reasons why the step in the process is a CCP.
- Think about process steps on the flow diagram in identifying CCPs, not just physical points in time and space.

Step 8: Establish Critical Limit(s)

For each CCP identified in the process, a measurable parameter and limits for each parameter must be established. Current knowledge and expertise including industry standards, as well as historical data, should be used as a guide when determining the limit.

To provide appropriate control of the water distribution system, a two-tiered structure may be implemented. The first level of critical limits distinguishes when the measurable control parameters are approaching an unacceptable level. The second level identifies the absolute cut-off value that ensures the quality and safety of the water will not be compromised. That is when the water becomes unacceptable and is considered off-specification or unfit and unsafe to drink.

A simpler way of thinking is the first set of critical limits, called target limits or target criteria, is the optimum values. The second set of critical limits is the operational limits that trigger actions to rectify the deviation and bring the water back within specification.

Table 3.14 provides a summary of examples of critical limits in utility HACCP plans and covers the range of different types of limits that can be set.

Table 3.14
Examples of critical limits in utility HACCP plans

Process, operation or procedure identified as a CCP	Examples of preventive measure associated with the CCP	Examples of critical limit	Utility source
Closed reservoir storage	Maintain integrity of storage including security devices and bird-proofing	• No breach in bird-proofing, roof, hatches, or security	Power and Water Corporation, Northern Territory, Australia
Chlorination on outlet of open reservoir	Maintain free residual to control bacteria from birds	• Free residual above 0.5 mg/L at entry point	Barwon Water, Victoria, Australia
Chlorination on inlet of closed reservoir	Maintain free residual to control ammonia oxidizing bacteria development	• Free residual above 0.2 mg/L at reservoir outlet • Free residual below 0.6 mg/L at first customer	Sydney Water, New South Wales, Australia
Maintenance of positive pressure in system	Maintain water in tanks to keep system pressurized	• To maintain pressure in system, reservoir levels never to drop below design service levels at any time • Specific levels are set for each tank to give ≥ 49 ft (15 m) head at all times	Power and Water Corporation, Northern Territory, Australia
	Maintain distribution system pressure	• Pressure should be above 35 psi (241 kPa) under normal conditions • Pressure should be above 20 psi (140 kPa) during emergency conditions	City of Austin, Texas.
Backflow prevention system	Presence of working backflow prevention device	• Compliance with plumbing code • Backflow assemblies pass operational test	City of Austin, Texas

The simplest example is chlorination levels. The water leaving an open service basin might be considered unacceptable if it has a residual chlorine level of less than 0.2 mg/L at the monitoring point leaving the contact tank. The disinfection process might be designed to produce a residual chlorine level of 0.5 mg/L at that point. If monitoring (see Step 9) detects that the water has a chlorine level of 0.25 mg/L, the dosage of sodium hypochlorite might be programmed to increase to raise the residual chlorine level in the system to 0.5 mg/L, the target level. If the level still drops and goes below 0.2 mg/L, which is the critical limit, a corrective action (see Step 10) might be taken, such as spot-dosing in the downstream service reservoir or shutting off the feed and running off the high level tank until the problem is fixed.

Critical limits can involve less quantitative criteria than those generally used for chlorine residuals. For example, observational measures, such as integrity of fencing or the presence/ absence of devices such as backflow prevention devices, have been identified as critical limits in some HACCP plans.

Hints:

- Document the basis for setting critical limits (see step 11, Validate/Verify HACCP Plan).
- Establish target values that detect when the water is trending in an unacceptable direction, as well as critical limits when it reaches an unacceptable value.
- Consider setting critical limit values for aesthetic reasons such as taste and odor, as well as health issues.

Step 9: Establish a System to Monitor Control of the CCP

Monitoring procedures under step 9 serve three main purposes:

- Ensure the process system is under control and determine when target criteria and critical limits are reached, i.e. when a deviation occurs at a critical control point;
- Provide records that can be later examined during the verification step; and
- Track the effectiveness of the operation over time.

It is vital that monitoring procedures are established and conducted properly. Points to be taken into account when establishing monitoring procedures include:

- What specific information is required to be collected?
- How is the data to be collected and recorded?
- Where should the monitoring be undertaken?
- Who is to undertake the monitoring?
- When should the monitoring occur, i.e. at what frequency?
- Whether further analysis is required of the raw data collected to enable a meaningful comparison to the critical limit?

The designated parameter(s) determined for each CCP is/are to be measured prior to the point where unsafe or poor quality water may reach consumers. This means that many parameters will need to be monitored online so that "real time" results are obtained as few "storage points" exist within a water distribution system. The monitoring procedures need to be documented in the HACCP plan (see examples in Table 3.15). Factors to be considered when determining appropriate monitoring are:

- Available monitoring methods;
- Costs for sample collection and analysis;
- Time required for sample collection and analysis; and

- Training requirements.

For example, *E. coli* is analyzed as a measure of the safety of the water. Results for *E. coli* however, take at least 24 hours from sampling before a confirmed result is returned. Unless the water remains in the pipes or tanks and is not consumed by customers, this parameter is not an effective monitoring control. Instead, such a parameter would be used for verification monitoring (under step 11) rather than operational monitoring. A more appropriate parameter for drinking water distribution systems may be chlorine residuals, pressure, tank levels, or turbidity.

Table 3.15
Examples of critical limit monitoring in utility HACCP plans

Critical limit	Example of monitoring	Utility
No breach in bird-proofing, roof, hatches or security on closed service reservoir	What: Integrity How: By observation Where: Reservoir Who: Field services staff When: Fortnightly (biweekly)	Power and Water Corporation, Northern Territory, Australia
Free residual above 0.5 mg/L at entry point to distribution system for water leaving open service reservoir	What: Chlorine residual How: Chlorine analyser Where: Entry point Who: Duty manager When: Online	Barwon Water, Victoria, Australia
Pressure should be above 35 psi (241 kPa) under normal conditions Pressure should be above 20 psi (140 kPa) during emergency conditions	What: Pressure How: Pressure transducer Where: Distribution system Who: Alarm monitor When: Online	City of Austin, Texas.
Location and type of backflow assemblies comply with Plumbing Code Backflow assemblies pass operational test	What: Device compliance How: Testing regulation Where: Service connection Who: Contract plumber When: Annually	City of Austin, Texas

At Gold Coast Water in Australia, improved monitoring and reporting procedures implemented as part of the HACCP Plan have provided more information to management and have motivated operational staff to work harder to avoid failures (Smith 2004). Operational staff members were consulted during development of the HACCP plan to identify realistic critical limits and monitoring procedures. When a critical limit is exceeded at Gold Coast Water, the operators submit an excursion report to management (Smith 2004). For example, a plant operator may file an excursion report because of failure to maintain treated water pH for 4 hours due to aging lime dosing equipment. Management can quickly review these excursion notices and decide if they represent minor incidents or if further inquiry is warranted. Operators do not like reporting excursions so they try harder to avoid having failures in the first place. It is hard to cover up failures because the utility has numerous on-line devices to measure water quality conditions (e.g. turbidity, pH, chlorine) and close internal auditing methods which that reveal cover ups (Smith 2004a).

At Gold Coast Water's Mudgeeraba plant, Smith (2003b) reports that there was initially under-reporting of turbidity problems due to problems with the sensitivity and calibration problems with the plant's only on-line turbidity meter (for 16 filters). This was compounded by a culture that was uncomfortable with and unused to communicating water quality data on a continuous basis. A change to better instrumentation and closer auditing eventually revealed that turbidity was not being controlled to the level of consistency expected. Two problems were revealed. First, the operator's skills in carrying out jar tests had deteriorated due to lack of practice and, second, problems with the filters and backwash system made turbidity failures more likely and of more serious consequence. The cultural problem was addressed through tighter auditing which eventually led to better understanding of the importance of the staff's role in managing water quality. The asset problem led to a redesign of the backwash system and an upgrade of the plant filters.

Monitoring strategies associated with a HACCP Plan do not rely solely on water quality testing, but may also include inspections and checks on various databases or other utility records. For example, as summarized in Appendix B, the City of Austin, Texas focused on two hazards for their HACCP Plan – cross-connections and new construction sites. Monitoring strategies for evaluating control measures for the cross connection hazard include:

- Plumbing inspections and water protection surveys to identify any unprotected cross-connections and to check for working backflow prevention devices.
- Plumbing inspections and water protection surveys to check on repairs of failed backflow prevention devices.
- Pressure monitoring throughout the distribution system to evaluate whether system pressure is being maintained.
- Reviews of the utility's database to check if annual inspections of backflow prevention devices were conducted.
- Visual inspection of on-site septic systems to check for system failures.

Monitoring strategies for Austin's new construction site hazard include:

- Phone and radio contact with site personnel to check on unauthorized use of system valves.
- Visual inspection of contractor disinfection practices on new water lines.
- Water quality samples to check coliform bacteria counts and free chlorine levels of water in new mains prior to being placed into service.
- Monitoring tank water level and pressure point alarm levels that would indicate if any valves were opened between two pressure zones.
- Site inspections to check if contractors utilized the One-Call system (a communication system that provides a toll-free number for contractors/designers to call facility owners prior to excavating to prevent damage to underground facilities and utility lines) to mark water utility lines.
- Reviews of sign-in sheets for training workshops to check whether site inspectors attended required training sessions.

Hints:

- Select a monitoring frequency that enables a protective response to be taken before poor quality water reaches consumers.
- Include the person responsible for the monitoring in the development of monitoring procedures to ensure that they are practical and understandable.
- Ensure procedures and work instructions exist for monitoring at each CCP.

Step 10: Establish Corrective Actions

For every monitoring procedure, corrective actions need to be developed for situations when target criteria and critical limits have been deviated from to ensure that the critical control point is brought under control or that unsafe water has been disposed of in a correct and appropriate manner. The corrective action procedures need to be tested, either by a desk-top study or field trial, to ensure that water that may be unfit or unsafe is prevented from reaching consumers.

Any and all corrective actions undertaken need to be recorded in the HACCP system as proof of compliance and to prove that all reasonable precautions have been taken in accordance with due diligence.

Examples of corrective actions for the distribution system include flushing or air-scouring, booster dosing of disinfectant, operating valves, removing a pipe from operation, redirecting flow, or taking a piece of equipment (tank or disinfection unit) offline and bypassing the asset until control is restored. Additional examples are provided in Table 3.16.

Table 3.16
Examples of corrective action procedures in utility HACCP plans

Critical limit	Example of corrective action	Utility
No breach in bird-proofing, roof, hatches or security on closed service reservoir	• Isolate/bypass storage • Spot-dose • Scour out suspect water • Clean and restore reservoir	Power and Water Corporation, Northern Territory, Australia
Free chlorine residual drops by 0.2 mg/L or more from the chlorination set-point for water leaving an open storage reservoir.	• Ensure telemetry system is operating correctly. • Review/adjust the set-point or increase the dosing level. • Initiate the contingency plan for chlorine plants.	South East Water, Victoria, Australia
Pressure should be above 35 psi (241 kPa)under normal conditions; and Pressure should be above 20 psi (140 kPa) during emergency conditions	• Turn on additional pumps to raise pressure • Open divide valves to charge system from different zone • Search for main breaks and shut off leaking area • Issue boil water advisory if necessary	City of Austin, Texas
Location and type of backflow assemblies comply with Plumbing Code Backflow assemblies pass operational test	• Remind customer • Enforce compliance using penalties if required	City of Austin, Texas

Corrective actions may not necessarily be carried out where the hazard event occurs but may be performed upstream or further downstream. Corrective actions in the water distribution system may be carried out because of a hazardous event that has occurred at the water treatment facility or in the watershed.

Hints:

- Note that some corrective actions may be the same as control measures, such as flushing.
- Document corrective actions and monitoring procedures next to the relevant CCP in the HACCP plan so that it can be easily followed when required.
- Identify the people responsible for undertaking corrective actions and ensure they have the required authority and training.
- Remember to include disposal or isolation and treatment of potentially unsafe water.
- Corrective actions are not always undertaken where the hazardous event occurs.
- Identify other parties that may be responsible for undertaking certain corrective actions.
- Review records of corrective actions to identify trends and the need for systematic improvement actions.
- Utilize emergency response plan and training materials to develop and implement corrective actions.

Step 11: Validate/Verify HACCP Plan

Step 11 involves two types of activities, validating the technical basis for the HACCP Plan, and verifying the HACCP Plan which confirms that the HACCP system has been accurately implemented and is working as designed.

Validate the HACCP Plan

Validation is a means of substantiating that the HACCP Plan is technically sound. For example, there is a need to ensure that the stated critical limits are correct and effective for producing and maintaining safe and potable water. For example, what is the proof that the value specified is correct and controls the hazard? (See example in Table 3.17.) Proof of the technical validity of the HACCP Plan needs to be documented in the plan and may make reference to sources such as:

- Technical literature and research reports;
- In-house validation exercises (i.e. specific sampling regimes to demonstrate what is occurring);
- Design specifications from reputable professional engineering firms;
- Regulatory requirements;
- Memoranda of understanding; or
- Guidelines and regulations produced by reputable professional bodies of experts.

Table 3.17
Example validation schedule

Critical limit	Validation
Reservoir levels never to drop below design service levels at any time	Distribution network modeling and pressure recording has verified that the minimum network pressure can be maintained when the storage levels are managed within operational limits.
Levels are set for each tank to give ≥ 49 ft (15 m) head at all times	The minimum network head of 49 ft (15 m) is consistent with best practice for preventing or mitigating pathogen intrusion problems e.g. *Recommended Standards for Water Works* (Ten State Standards) 1997 nominate the minimum supply pressure as 20 psi (140 kPa).
Free chlorine residual in the reticulation system should be between 0.2 and 0.5 mg/L	To preserve the microbial quality of water during distribution the Australian Drinking Water Guidelines (2004) recommend a chlorine residual of ≥ 0.2 mg/L in the reticulation system* but below 0.6 mg/L to control taste and odor complaints.

*Note that although this is not essential in much of the USA and Europe, in the Northern Territory of Australia, water can become very warm (>30°C (86°F)) and pathogens such as *Nagleria fowleri* and *Burkholderia pseudomallei* can become problematic.

Source: Adapted from Power and Water Corporation's HACCP Plan (Appendix B)

Verify the HACCP Plan

This step determines that all hazards have been identified and that CCPs, critical limits, monitoring and corrective action procedures are appropriate and effective. The verification process entails obtaining objective evidence that the water is, in fact, safe and fit to drink and that good operational practices, monitoring and corrective actions are being complied with at all levels. This step is achieved by means such as:

- Conducting regular internal audits to ensure that utility staff are following SOPs;
- Revisit entire HACCP Plan
- Cooperate with audit conducted by independent third party and correct any noted deficiencies.
-

Verification activities should be completed according to a predetermined schedule as developed by the HACCP team, as illustrated by the example in Table 3.18. At South East Water in Victoria, Australia, internal verification procedures include the following (Mullenger 2002):

- Random water quality sampling to substantiate that the water is safe;
- Daily check of exceedance records such as indicator organisms and customer complaints; and
- Checks on appropriate corrective actions (booster chlorine dosing or flushing).

Table 3.18
Example verification schedule

Activity	Description	Frequency	Responsibility	Records
Water quality monitoring	Customer tap sites throughout the supply system are sampled on a regular basis by an accredited independent Laboratory. This sampling program verifies the effectiveness of the HACCP System. Exceedances are monitored daily and field responses and re-sampling undertaken accordingly. Automated procedure in place for processing the daily exceedance notifications from the testing laboratory and generation of field response requests to the maintenance contractor.	Monthly Reference: annual water quality monitoring program	Manager Water Quality Planning	• Water quality database • Automated exceedance management SOP • Exceedance database • Department of Health Report • Monthly water quality report
Audit HACCP Plan	Internal and external audits are conducted on a regular basis to identify areas of poor performance and opportunities for improvement of HACCP plan. Audits also verify that activities comply with documented requirements.	Annual Reference: Management System Core Process	Business Improvement Coordinator	• Audit Schedule • Audit Reports • Management system core process SOP
Review staff training	• All staff in critical areas of process operation and control is regularly trained. • Relevant staff are trained annually by an accredited HACCP Consultant • Relevant staff are annually trained by Water Industry Training Centre	Annual Reference: Management System - Core Process – Manage Human Resources	Manager Water Quality Planning	• Computer based personnel system "Human Resource Information System" • Performance plans • Training manual

Source: adapted from Yarra Valley Water 2005. Reprinted with permission.

The HACCP system should be revisited in its entirety by the HACCP team at least annually, when a hazardous event has occurred, or a significant system change has taken place. For example, the Plan is may be reviewed in the following circumstances (Mullenger 2002):

- Changes to water quality zone boundaries or the disinfection process;
- Equipment or facility modifications such as covering an open reservoir;
- Regulatory amendments;
- Changes in piping materials or repair techniques; and
- Occurrence of a significant hazardous event such as a security breach.

Water quality monitoring is often conducted as part of the external audit to verify that the HACCP plan is effective at producing water without unsafe or aesthetically unacceptable levels of hazardous contaminants. This type of verification monitoring may also be conducted by the utility.

The frequency at which verification auditing and monitoring is carried out depends on how long the HACCP team feels the system can remain unchecked before deviations might go undetected. Some items may require checking on a daily basis, while others may only need to be confirmed annually. For example, checks of coliform bacteria and customer complaints may occur daily, while audits on the emergency and maintenance work occurring within the water distribution supply system may occur on a weekly basis. Review of corrective actions may be scheduled on a monthly basis and supporting programs may be reviewed annually.

Hints:

- Document verification schedules (i.e. items and their frequencies) within the HACCP plan.
- Review monitoring and corrective action procedures.
- Regularly review data trends to identify areas for improvement.
- Check on calibration schedules and records to assure that calibration is being accomplished.
- Include the person(s) responsible for each verification activity in the HACCP plan.

Step 12: Establish Documentation and Record Keeping

"Efficient and accurate record keeping is essential to the application of a HACCP system" (Codex Alimentarius Commission 1993). It provides retrospective proof of compliance to the HACCP plan and enables product traceability and facilitates continuous improvement from trend analysis. Record keeping can even support a legal defense. All HACCP documents and records should be dated and signed (Mullenger 2002).

Documentation and record keeping should be appropriate to the nature and the size of the organization operating the water distribution system. Examples of record-keeping and documentation for the distribution system include:

- All information used to develop the HACCP Plan;
- Regulatory requirements;
- Utility water quality and operational goals for distribution system;
- Water quality monitoring program;
- Monitoring results including deviations;
- Associated corrective actions undertaken when a deviation occurs;
- Verification activities (e.g. audit reports);
- Validation records (e.g. EPA regulations and guideline);
- Modifications and reviews of the HACCP system; and
- Employee training and competency records.

A detailed example of CCP records for one distribution system process step is provided in Table 3.19.

Table 3.19
Example of critical control point records for one process

Process (owner)	Causal events controlled	Hazards controlled
Pressurization of the distribution system (Water Distribution Engineer)	Re-contamination through ingress from loss of pressure leading to suction (**recorded** in risk register).	Microbial, chemical and physical hazards (**recorded** in risk register)
Critical limits	**Monitoring**	**Corrective Actions**
Tank level for reservoir-pressurized systems never below low-low alarm level at any time (actual level different for each tank, **recorded** in the SCADA telemetry alarm system).	What: Tank level Who: Telemetry with alarms to Hydraulics Duty Officer oversight by Water Distribution Engineer. Where: In each service reservoir with some in the distribution system. When: On line continuous. **Records**: CITEC	What: Turning on additional pumps to maintain pressure. Open valves to re route water. Flush lines if negative pressure occurred. Incident management plan if not rapidly controlled. Who: Hydraulics Duty Officer and Field Services Hydraulics Workers. Where: On site. When: Action initiated within 15 minutes of deviation alarm. **Records**: Corrective Actions Register.
Pressure from pump-pressurized systems never below low pressure alarm setpoint of $\geq$ 49 ft (15 m) head at lowest point (actual level different for each system, **recorded** in the SCADA telemetry alarm system).	What: Pressure Who: Telemetry with alarms to Hydraulics Duty Officer oversight by Water Distribution Engineer. Where: At the pressure monitoring SCADA point of each pumped system. When: On line continuous. **Records**: CITEC	
Validation	*Verification*	*Supporting Programs*
Hydraulic design specifications are based on ensuring pump pressures and tank levels will pressurize the entire system for adequate flow to meet service standards to give $\geq$ 15 m head at all times (**record:** Water Supply and Sewerage Standards Code). Pressure tests were performed at peak day demand at lowest pressure points in system to validate this (**recorded** in LIMS).	Annual review by Water Distribution Engineer to inspect records of monitoring, corrective actions and calibration (**recorded** in Verification Schedule). Results from pressure and flow meters at key points are checked to see that pressure remains positive at all times (**recorded** in CITEC). Customer complaints related to pressure checked to ensure they are not caused by systematic pressure drops (**recorded** in customer complaints database). Compliance static pressure monitoring by independent lab at distal sites monthly (**recorded** in LIMS).	Calibration and maintenance of flow, pressure, and level gauges as directed by Maintenance Manager, Water twice per year (**recorded** in maintenance tracking system). Compliance with flow and pressure design practices (**recorded** in Water Supply and Sewerage Standards Code). Staggered watering during hot days (**recorded** in water restriction SOP).

Source: Adapted from the ActewAGL 2005 with permission.

Hints:

- Maintain only the current version of documents to prevent duplication or confusion.
- Sign and date key documents to demonstrate that they are approved for use.
- Document the hazard analysis and CCP determination including critical limits.
- Consider using checklists and flow diagrams to simplify record-keeping.
- Record monitoring results, deviations, corrective actions and reviews.
- Review records for trends to identify the need for preventative action.
- Develop the documentation and record keeping system with the personnel that will be ultimately using them.
- Maximize the use of electronic and simple record keeping in order minimizing paperwork.

CHAPTER 4
PILOT TESTING OF HACCP PLANS

INTRODUCTION

This chapter describes pilot-testing activities conducted by four participating utilities, including Sydney Water Corporation in Australia, Power and Water Corporation in Australia, the City of Austin, Texas, and the South Berwick Water District in Maine. The locations of the Australian pilot study utilities are shown in Figure 4.1. The objective of the pilot testing is to develop HACCP Plans based on the tailored HACCP system developed as part of this Project and described in Chapter 3, and then implement the Plan for a 12-month testing period. The results of the implementation experiences are used along with utility input to assess the value of the HACCP system for protecting and maintaining distribution system water quality.

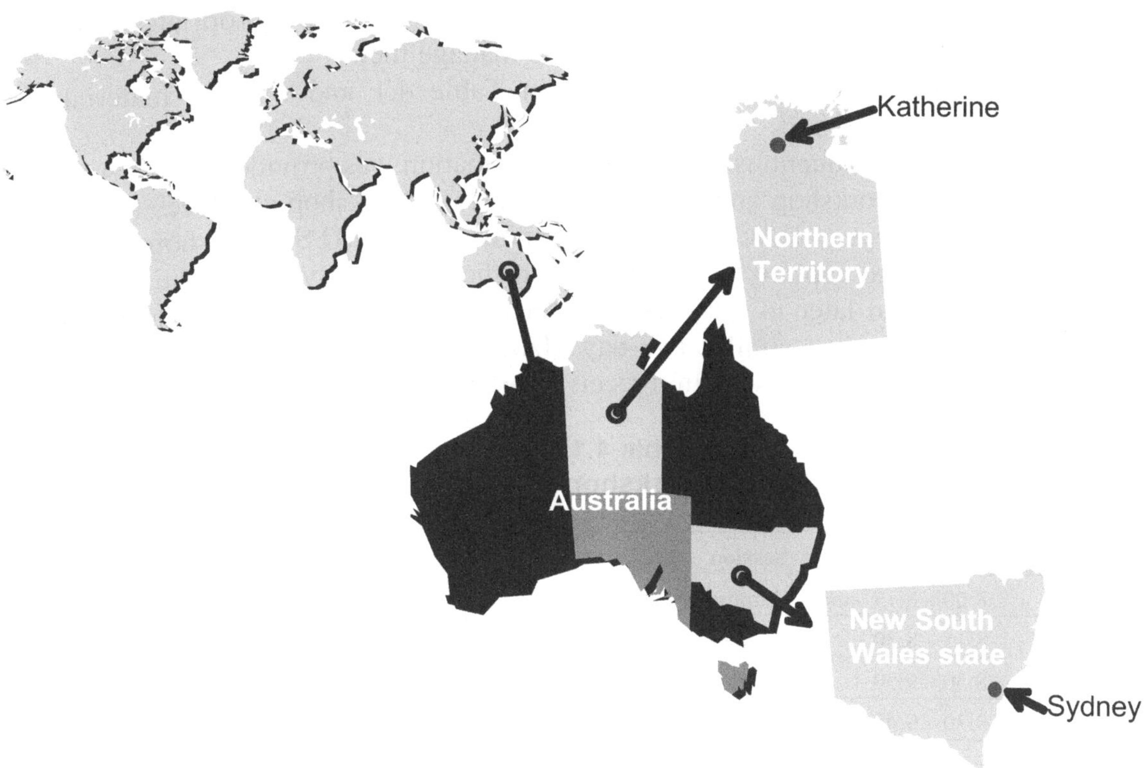

Figure 4.1 Map illustrating the relative positions of the Australian pilot study sites

APPROACH

For each participating utility, proposed pilot-study activities included a training workshop, formation of a HACCP team, development of a HACCP Plan, implementation of the Plan, and preparation of a final report. The Project Team conducted the training workshops and assisted the utility's HACCP team with aspects of the HACCP Plan development and implementation activities. The Project Team also developed surveys and criteria by which the utilities could assess their experiences with HACCP. The approach used for these pilot-study activities is discussed in the following sections.

Training Workshops

A training workshop was held at each utility location so that the Project Team could train utility staff on the HACCP principles and initiate development of the utility's HACCP plan. Invitations to attend the workshop were extended to utility staff with a broad range of expertise and skill in all areas related to distribution system water quality (e.g. management, laboratory, operations, engineering, and construction). The workshop's success depended on the caliber of workshop attendees; therefore, it was essential to include individuals with appropriate knowledge of the potential hazards and the control measures used to manage them.

An example workshop agenda is presented in Table 4.1 and training materials are included in Appendix A to this report.

Following informal introductions, background information was provided on the AwwaRF project, the purpose of the workshop and HACCP principles. Workshop attendees then began working through the basic steps of HACCP. At the conclusion of the U.S. workshops, attendees completed a written survey to capture their opinions on the HACCP system. The survey and survey findings are discussed later in this chapter. Such a survey was not completed for the Australian workshops, although a similar two-day HACCP workshop conducted as an open workshop for the Australian Water Association received high scores.

Table 4.1
Typical Workshop Agenda

Time	Session
8:00 – 8:15	Purpose of Workshop
8:15 – 8:30	AwwaRF Project Overview
8:30 – 9:30	What is HACCP?
9:30 – 9:45	Break
9:45 – 10:45	Small Groups, Steps 1-5
10:45 – 11:45	Small Groups, HACCP Step 6
11:45 – 12:00	Wrap-Up, Small Group Findings
12:00 – 1:00	Lunch
1:00 – 2:00	Small Groups, HACCP Step 7
2:00 – 3:00	Small Groups, HACCP Steps 8, 9, and 10
3:00 – 4:00	Small Groups, HACCP Steps 11 and 12
4:00 – 4:30	Workshop Wrap-Up

Formation of HACCP Team

Each participating utility formed a HACCP team to complete the HACCP plan and to guide its implementation. Ideally, team members were a subset of the utility staff that attended the workshop. As with the workshop, it was important to include utility management on the HACCP team as well as utility staff with varying expertise and responsibilities related to distribution system water quality.

Development of HACCP Plan

Depending on the amount of work accomplished at the training workshop, the HACCP team then completed the HACCP Plan in one or more work meetings. The Team also consulted with various utility staff to confirm the accuracy of the process flow diagram and the classification of hazards, and to collect documentation on utility standard operating procedures, monitoring and inspection programs, and data management systems. HACCP Plans for Austin, South Berwick, and Power and Water Corporation are provided in Appendix B and are discussed in the section below entitled "Pilot Study Results." The Sydney Water HACCP plan is not included in Appendix B because it is in a preliminary draft form. Management plans developed as part of the HACCP process are in Appendix D.

Development of Surveys and Evaluation Criteria

The survey shown in Figure 4.2 was developed to capture utility feedback on the HACCP system following the U.S. training workshops. Survey findings are discussed later in this chapter for each utility.

Dear Workshop Participant,

Thanks for taking the time from your regular work to participate in the HACCP Workshop on June 25th. As part of our research, we would like your specific feedback on the workshop to determine if this HACCP process is useful to other utilities. Please take a few minutes and respond to the following questions. Thank you!

1. At the workshop, did you gain new information about South Berwick's distribution system management practices or potential hazards?
Circle One: Yes No Maybe
If yes, please summarize.

2. Do you think other workshop participants better understand your area of responsibility and/or water quality concerns as a result of the workshop?
Circle One: Yes No Maybe

3. After the workshop, did you think of any additional hazards or critical control points that were not mentioned that day? If so, please list here.

4. Would you suggest the HACCP process to other utilities for improving distribution system protection?
Circle One: Yes No Maybe

5. Would you like to see the South Berwick Water District apply the HACCP approach to to the whole system, source to tap?
Circle One: Yes No Maybe

6. Please comment on the workshop structure:

a. Length of time: (Circle One) – too long too short ok

b. Size of group (Circle One) – too large too small ok

c. Workshop notebook content (Circle One):
too detailed not enough detail ok as is

d. Presentation content (Circle One):
too detailed not enough detail ok as is

Additional Comments:

Thanks for your input!

Figure 4.2 Example workshop survey form for South Berwick Water District

A set of evaluation criteria was used to assess the success of pilot testing. This evaluation was completed as a written survey by the utility HACCP team. The evaluation criteria are summarized in Table 4.2.

Table 4.2
Evaluation criteria for pilot studies

Question	Quantitative or qualitative	How assessed?	When?	By whom?	Answer*	Basis/ examples
Group 1. Has HACCP implementation improved customer satisfaction?						
• Reduced taste/odor complaints	Quantity	Utility records	At end of monitoring period, compared with the start.	HACCP team		
• Reduced discolored water complaints	Quantity	Utility records	At end of monitoring period, compared with the start.	HACCP team		
• Reduced low pressure complaints	Quantity	Utility records	At end of monitoring period, compared with the start.	HACCP team		
Group 2. Has HACCP improved regulatory compliance?						
• *E. coli* or thermotolerant coliforms?	Quantity	Utility records	At end of monitoring period, compared with the start.	HACCP team		
• Total coliforms?	Quantity	Utility records	At end of monitoring period, compared with the start.	HACCP team		
• HPCs?	Quantity	Utility records	At end of monitoring period, compared with the start.	HACCP team		
• Turbidity?	Quantity	Utility records	At end of monitoring period, compared with the start.	HACCP team		
• Low pressures?	Quantity	Utility records	At end of monitoring period, compared with the start.	HACCP team		
Group 3. Has HACCP provided stronger mechanisms for managing risks to public health than previously?						
• Reduced water quality incidents[†]?	Quantity	Utility records	At end of monitoring period, compared with the start.	HACCP team		

(Continued on next page)

Table 4.2 (Continued)
Evaluation criteria for pilot studies

Question	Quantitative or qualitative	How assessed?	When?	By whom?	Answer*	Basis/ examples
• Reduced chlorinator failures?	Quantity	Utility records	At end of monitoring period, compared with the start.	HACCP team		
• Systematic identification and prioritization of risks and control?	Quality	Judgment by water quality s[illegible]	At end of monitoring [illegible] compared he start.	HACCP team		
• Improved understanding of the distribution system?	Quality	Judgment b[illegible] quality s[illegible]	[illegible] monitoring compared he start.	HACCP team		
• Improved understanding and maintenance of positive system pressure?	Quality	Judgment b[illegible] quality s[illegible]	[illegible] monitoring compared with the start.	HACCP team		
• Enhanced backflow and cross-connection prevention?	Quality	Judgment by water quality staff	At end of monitoring period, compared with the start.	HACCP team		
• Has monitoring become more focused?	Quality	Judgment by water quality staff	At end of monitoring period, compared with the start.	HACCP team		
Group 4. Has HACCP improved system management and control?						
• Improved databases?	Quality	Judgment by management	At end of monitoring period, compared with the start.	HACCP team		
• Improved calibration schedules?	Quality	Judgment by management	At end of monitoring period, compared with the start.	HACCP team		
• Better targeted reporting?	Quality	Judgment by management	At end of monitoring period, compared with the start.	HACCP team		
• Improved record keeping?	Quality	Judgment by management	At end of monitoring period, compared with the start.	HACCP team		
• Made it easier for O&M staff to operate system?	Quality	Judgment by management	At end of monitoring period, compared with the start.	HACCP team		

(Continued on next page)

Table 4.2 (Continued)
Evaluation criteria for pilot studies

Question	Quantitative or qualitative	How assessed?	When?	By whom?	Answer*	Basis/ examples
Group 5. Has HACCP improved human factors?						
• Enhanced operator empowerment (ownership, understanding and involvement)?	Quality	Judgment by Human Resources and/or management	At end of monitoring period, compared with the start.	HACCP team		
• Did additional training in water safety get introduced?	Quality	Judgment by Human Resources and/or management	At end of monitoring period, compared with the start.	HACCP team		
• Did existing training become less ad-hoc and more structured?	Quality	Judgment by Human Resources and/or management	At end of monitoring period, compared with the start.	HACCP team		
• Improved senior management involvement?	Quality	Judgment by Human Resources and/or management	At end of monitoring period, compared with the start.	HACCP team		
• Improved relationships between management and operators?	Quality	Judgment by Human Resources and/or management	At end of monitoring period, compared with the start.	HACCP team		
• Improved clarity of accountability?	Quality	Judgment by Human Resources and/or management	At end of monitoring period, compared with the start.	HACCP team		
• Has communication been improved within the utility and between the utility and external stakeholders?	Quality	Judgment by Human Resources and/or management	At end of monitoring period, compared with the start.	HACCP team		

(Continued on next page)

Table 4.2 (Continued)
Evaluation criteria for pilot studies

Question	Quantitative or qualitative	How assessed?	When?	By whom?	Answer*	Basis/ examples
Group 6. Costs and benefits of HACCP.						
• What are the estimated costs of implementation of the HACCP system?	Quantity	Best estimate	At start and end of monitoring period	HACCP team		*e.g. How many meetings in-house, paid support, people time in–house on the project.*
• What new processes and concepts did HACCP bring in?	Quality	Judgment by utility HACCP Team	At start and end of monitoring period	HACCP team		*e.g. internal and external auditing*
• Was good use made of existing systems to minimize duplication and maximize integration?	Quality	Judgment by utility HACCP Team	At start and end of monitoring period	HACCP team		
• What is the estimated value of the benefits of these changes?	Quantity	Best estimate	At end of monitoring period, compared with the start.	HACCP team		*e.g. reduced disinfection and monitoring costs?*
• Has HACCP enhanced any environmental outcomes?	Quality	Judgment by utility HACCP Team	At end of monitoring period, compared with the start.	HACCP team		*e.g. discharge of chlorinated water.*
• How was the auditing (inspection, evaluation) process perceived?	Quality	Judgment by utility HACCP Team	At end of monitoring period, compared with the start.	HACCP team		
• Has continuous improvement been introduced?	Quality	Judgment by utility HACCP Team	At end of monitoring period, compared with the start.	HACCP team		
• Has credibility and due diligence improved?	Quality	Judgment by utility HACCP Team	At end of monitoring period, compared with the start.	HACCP team		

* To be completed by the project team in liaison with the utilities.
† Major water quality problems leading to some kind of emergency response such as extensive dirty water events, taste and odor events, or suspected contamination events.

IMPLEMENTATION OF HACCP PLAN

Each utility planned to implement their HACCP plan over a 12-month period during which certain operational and water quality parameters would be monitored. Monitoring data can consist of the following information:

- Data to show that critical limits at critical control points were within acceptable limits (e.g. chlorination records, set points);
- Documentation of record-keeping (e.g. staff training, chemical ordering, laboratory records);
- Water quality monitoring data for verification (usually end-point testing);
- Inspection reports; and
- Records of any emergency response, incident reporting and corrective actions.

The participating utilities used standard industry practices for sample collection and shipping, and calibration and use of field instruments. Where available, certified laboratories were used for laboratory analyses. These laboratories employed quality assurance procedures and audits as required to maintain certification status. The Project Team reviewed sampling program details with utility personnel prior to the start of monitoring.

PILOT STUDY RESULTS

South Berwick Water District

The South Berwick Water District in Maine, U.S., serves approximately 4,000 people with groundwater supplies including well points, a gravel well and drilled bedrock wells in three separate aquifers. The water is treated with chlorine and UV disinfection. The District maintains and services a distribution system that includes over 31 miles (50 km) of water main and a 1 MG (4 ML) concrete reservoir. The District is noted for its proactive management of the water system. The District has an active wellhead protection program for its many groundwater supplies, and regularly completes system maintenance such as biannual main flushing. The District has participated in several AwwaRF-funded research projects.

Training Workshop

A one-day HACCP training workshop was held at the South Berwick Water District office on June 25, 2003. Workshop attendees included the four staff members of the water district plus several outside experts needed to provide expertise on relevant topics. The outside experts are listed in Table 4.3. The workshop agenda was similar to the example provided in Table 4.1. Meeting discussions were held as one large group rather than small breakout groups. Workshop attendees were able to complete HACCP Steps 1 through 7.

Table 4.3
Outside experts invited to South Berwick workshop

Name	Organization	Title	Expertise
Terry Trott	State of Maine Drinking Water Program	Operator Licensing Officer	State regulations, cross connection control programs
J. Kevin Reilly	US EPA Region 1	Microbiologist	Microbiological hazards
Kathy Martel	EES, Inc.	Project Manager	HACCP, distribution system best management practices
Mark Arenberg	South Berwick Water District	Trustee	Engineering
Jeff Musich	Wright-Pierce	Vice President	Engineering

The results of the written survey conducted at the end of the training workshop are summarized in Figure 4.3. Seven of nine workshop participants completed the survey. The majority of survey respondents indicated that they gained new information about the distribution system during the workshop. All survey respondents thought that others gained a better understanding of their responsibilities and water quality concerns.

June 25, 2003

1. At the workshop, did you gain new information about South Berwick's distribution system management practices or potential hazards?

Circle One: Yes - 4/7 No – 2/7 Maybe – 1/7

If yes, please summarize. New information included status of current backflow prevention program and other O&M issues.

2. Do you think other workshop participants better understand your area of responsibility and/or water quality concerns as a result of the workshop?

Circle One: Yes – 7/7 No Maybe

3. After the workshop, did you think of any additional hazards or critical control points that were not mentioned that day? If so, please list here (None – 7/7)

4. Would you suggest the HACCP process to other utilities for improving distribution system protection?

Circle One: Yes – 6/7 No – 0/7 Maybe – 1/7

5. Would you like to see the South Berwick Water District apply the HACCP approach to to the whole system, source to tap?

Circle One: Yes – 6/7 No – 0/7 Maybe – 1/7

6. Please comment on the workshop structure:

e. Length of time: (Circle One) – too long 0/7 too short 0/7 ok 7/7

f. Size of group (Circle One) – too large 0/7 too small 0/7 ok 7/7

g. Workshop notebook content (Circle One):
too detailed 0/7 not enough detail 2/7 ok as is 5/7

h. Presentation content (Circle One):
too detailed 0/7 not enough detail 0/7 ok as is 7/7

Additional Comments:

1. Brought a better understanding of how HACCP can be applied to a small system like ours.
2. I am looking forward to see how this will work in Town. This will at least form a template for other small systems to utilize.
3. Well done – good process for rethinking.

Figure 4.3 Survey results for South Berwick Water District

Formation of HACCP Team

The HACCP team included the four members of the Water District staff including the superintendent, foreman, service person, and office manager.

HACCP Plan Development

A meeting was held September 8, 2003 to finish the draft HACCP Plan. The Plan was then submitted to the PAC and Australian team members for review. After incorporating review comments, the HACCP Plan was completed in November 2003. It is included in Appendix B.

The process flow diagram developed for the HACCP Plan is shown in Figure 4.4. Although the AwwaRF project focuses on HACCP as applied to water distribution systems, a decision was made at the South Berwick workshop to include the whole system source-to-tap in the HACCP Plan.

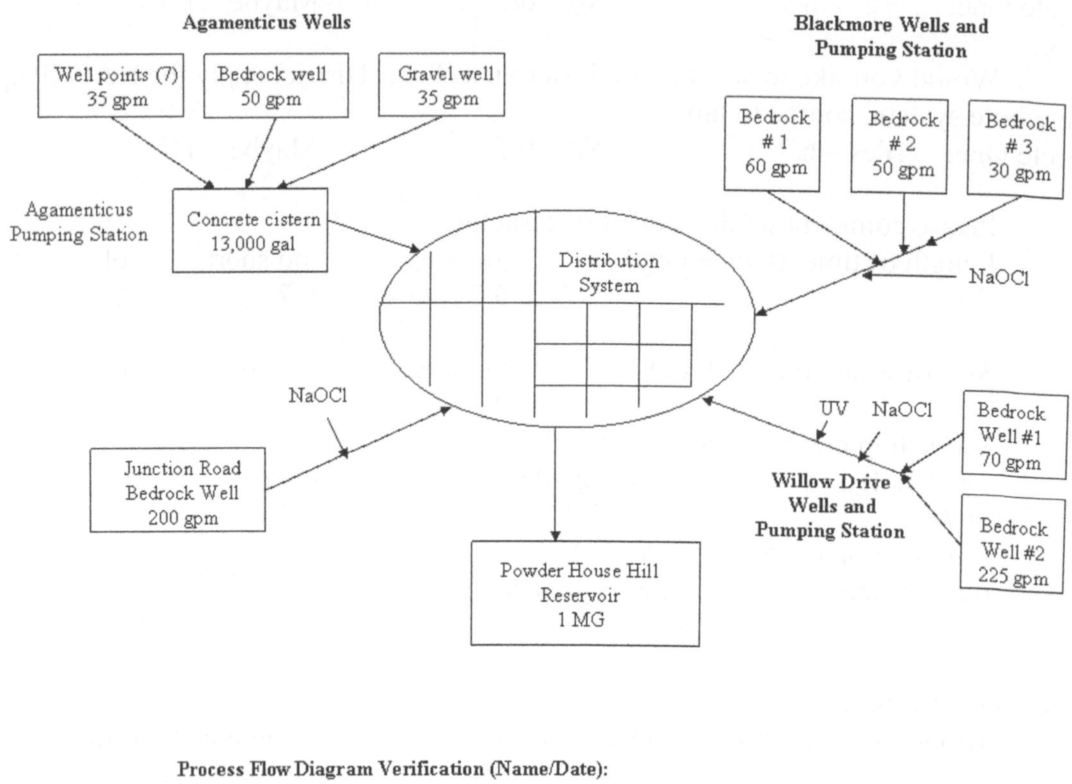

Figure 4.4 South Berwick Water District process flow diagram

In the hazard analysis (HACCP Step 6), 14 hazards were identified for South Berwick. A simple, semi-quantitative scoring approach, illustrated in Table 3.9, was used to rate each hazard. In this approach, a score is assigned to each hazard for two factors, the hazard's likelihood of occurring and the severity of consequences that could result from an occurrence (Gray and

Morain 2000; Deere et al. 2001). The scoring is based on a 1 to 5 scale. For the likelihood of occurrence factor, a score of 1 is given for a rare occurrence, while a score of 5 indicates a hazard that is almost certain to occur. Each hazard is also scored on the possible severity of consequences that could result, with a score of 1 for insignificant consequences and a score of 5 for catastrophic consequences. These two scores are multiplied to calculate the risk factor rating.

Eight of the fourteen identified hazards were classified as a low risk for South Berwick Water District. At the workshop, it was decided that a low risk would be defined as a risk score of 5 or less. These hazards include:

- Excessive water age in storage facility;
- Excessive water age in distribution system piping;
- Unintentional contamination of Willow Drive well from closed landfill or golf course;
- Installation of residential services;
- New construction;
- Contamination via pressure transients in area of former Superfund site/low pressure area;
- Vandalism/chemical contamination of Agamenticus well field; and
- Installation of service lines for fire protection.

Four hazards were determined to be a medium to high risk with risk scores ranging from 10 to 20. These hazards include:

- Backflow through an unprotected cross-connection (risk score 20);
- Intentional contamination via storage facility vents (risk score 10);
- Contamination due to service line break (risk score 10); and
- Contamination due to main break (risk score 10).

Two hazards were not classified due to a lack of water quality data. These hazards include:

- Unintentional contamination of shallow well points at Agamenticus well field; and
- Long dead-end mains with zero or poor disinfectant residual.

Due to limited resources, the District decided to focus on three hazards for its pilot study:

- Backflow through an unprotected cross-connection (risk score 20);
- Unintentional contamination of shallow well points (unrated); and
- Long dead-end mains with zero or poor disinfectant residual (unrated).

The focus of the District's HACCP Plan was to collect additional information to evaluate, document and improve control over these hazards. For example, although the District has a backflow prevention program, they lack complete records on the maintenance of backflow prevention devices owned by commercial customers.

At the time of the HACCP workshop (June 2003), the groundwater supplies at the Agamenticus wellfield were not disinfected. The District initiated disinfection in August 2003 at this wellfield. The initial idea for controlling the hazard of unintentional contamination at the

Agamenticus shallow well points was to conduct intensive monitoring for microbiological, chemical, and physical parameters in order to better characterize the microbiological risk. Per comments from the Project Team and with the advent of disinfection, the control of this hazard was changed to focus on the effectiveness of disinfection practices.

Implementation of HACCP Plan

Due to unforeseen circumstances, the South Berwick Water District withdrew from the Project in July 2004 without implementing their HACCP Plan. Nadeau (2004) explained that the Water District staff members were unable to implement the Plan due to other priorities, such as operating and maintaining the water system, building a new treatment facility, developing a new rate structure, dealing with local and state political issues, and struggling with unanticipated personnel and medical issues. South Berwick's experience illustrates the need for sufficient manpower to successfully implement a HACCP Plan.

City of Austin, Texas

The City of Austin's source of supply is the Colorado River. Untreated water for the pilot-study area is currently diverted at Lake Austin and treated at the Ullrich Water Treatment Plant using lime softening, recarbonation/pH adjustment, chloramination, filtration, and addition of ferric sulfate, fluoride, and sodium hexametaphosphate. Treated water is stored at the Ullrich Plant in two 10 MG (38 ML) clearwells and in the distribution system in tanks. The City owns and operates a contiguous distribution system that serves a population of approximately 770,000 through roughly 183,000 service connections. The distribution system contains 2,995 miles (4,819 km) of water mains of a wide variety of materials, including cast iron, ductile iron, PVC, asbestos cement, and reinforced concrete cylinder. The distribution system also contains 30 tanks ranging in size from 300,000 gallons (1 ML) to 34 MG (129 ML). Because of the varied topography in the City's service area, the distribution system is divided into eight major pressure zones.

Training Workshop

On May 21, 2003, the City held a one-day HACCP workshop. Attendees included staff with a broad array of skill sets from various divisions within the Utility with varying perspectives on distribution system operations. In particular, attendees included individuals from the Water Laboratory, Systems Planning, Cross-Connection Control, Process Engineering, Distribution System Operation, Water Quality, Regulatory Compliance, and the state's regulatory agency. The workshop agenda is shown in Table 4.1. Workshop discussions were held as one large group rather than small breakout groups. Workshop attendees were able to complete HACCP Steps 1 through 7 with the exception of Step 5 (Confirm Flow Diagram).

The results of the written survey conducted at the end of the training workshop are summarized in Figure 4.5. Nine workshop participants completed the written survey. All survey respondents indicated that they gained new information about the distribution system during the workshop. Most survey respondents thought that others gained a better understanding of their responsibilities and water quality concerns.

Formation of HACCP Team

Following the workshop, the Utility pulled together a multidisciplinary team to develop, verify and implement a HACCP plan. Team members are shown in Table 4.4.

Table 4.4
Austin's HACCP team

Name	Position	Role and responsibility
Barrios, Rosie	Water Laboratory Supervisor	Laboratory and analysis expert
Bennett, Tony	TCEQ Regulatory Manager	State regulator
Bohr, Onnie	Infrastructure Superintendent	Field operations and maintenance expert
Burazer, Jane	Asst. Director of Treatment	Treatment expert
Kuhn, Robert	Cross Connection Control Supervisor	Cross connection expert
Lutes, Teresa	Engineer/Planner	Systems planning expert
Ojeda, Edward	Construction Inspector	Construction inspection expert
Pedersen, Dan	Water Quality Manager	Team leader

May 2003

1. At the workshop, did you gain new information about Austin's distribution system management practices or potential hazards?

Circle One: Yes – 9/9 No – 0/9 Maybe – 0/9

If yes, please summarize.

a. The most useful part of the workshop was learning more about system operations, and dialogue with other groups about how they do their jobs.

b. I learned how complicated the distribution system is and how difficult it is to manage.

c. The City of Austin seems to be very aware of potential hazards, and seems to have adequate management already in place.

d. The ways in which the other divisions do their job and the importance of all parties know how to properly identify hazards.

2. Do you think other workshop participants better understand your area of responsibility and/or water quality concerns as a result of the workshop?

Circle One: Yes – 7/9 No – 1/9 Maybe – 1/9

3. After the workshop, did you think of any additional hazards or critical control points that were not mentioned that day? If so, please list here (NONE – 9/9)

4. Would you suggest the HACCP process to other utilities for improving distribution system protection?

Circle One: Yes – 6/9 No – 0/9 Maybe – 3/9

5. Would you like to see the City of Austin apply the HACCP approach to the whole distribution system?

Circle One: Yes – 4/9 No Maybe – 5/9

One survey respondent said yes but indicated a concern on HACCP's impact on work processes.

6. Would you like to see the City of Austin apply the HACCP approach to the whole system, source to tap?

Circle One	Yes	No	Maybe
	3/9	0/9	6/9

7. Please comment on the workshop structure:

i.	Length of time: (Circle One) –	too long, 1/9	too short, 1/9	ok 7/9
j.	Size of group (Circle One) –	too large, 0/9	too small, 0/9	ok 9/9

k. Workshop notebook content (Circle One):

too detailed	not enough detail	ok as is
0/9	0/9	9/9

Presentation content (Circle One):

l.

too detailed	not enough detail	ok as is
0/9	2/9	7/9

Additional Comments:

1. The process of getting information out of participants and up onto the flip chart for recording for everyone to see was slow. Perhaps some better procedure for getting brainstorming ideas out and visible for other participants to see might be found.

2. HACCP seems to be a good system; however, I don't know if it will contribute that much to a city like Austin that seems to have covered all the bases already. A city that is not as well organized may benefit more from HACCP implementation.

Figure 4.5 City of Austin workshop survey results

HACCP Plan Development

The HACCP Team met several times during the summer of 2003 to finalize the remaining steps of the plan. The draft HACCP Plan was reviewed by the Project Team and the PAC during the Fall 2003. The HACCP Plan is included in Appendix B.

The process flow diagram developed for the HACCP Plan is shown in Figure 4.6. The utility selected one pressure zone in the distribution system to conduct its study. The Southwest C Pressure Zone is located on the periphery of the Utility's distribution system and was selected because of a number of boil water advisories issued within that zone in recent years. Additionally, the Southwest C Pressure Zone has other features that make it appropriate for a HACCP plan. It contains four routine monitoring sites for the Total Coliform Rule that provide a history of water quality in the area. One of the monitoring sites also serves as a monitoring point for compliance with the Disinfection By-Products Rule, which further enhances understanding of water quality in the area. The Southwest C Pressure Zone is located on the extreme southwestern edge of the distribution system, containing maximum water age for that portion of the distribution system. Water traveling to this area passes through two tanks and three pump stations, which have the potential to affect water quality by increasing water age. It is a growing area with new construction occurring. It is a semi-rural area with older homes on septic systems that, should they fail, might pose a hazard to leaking water mains subject to low or negative pressure transients.

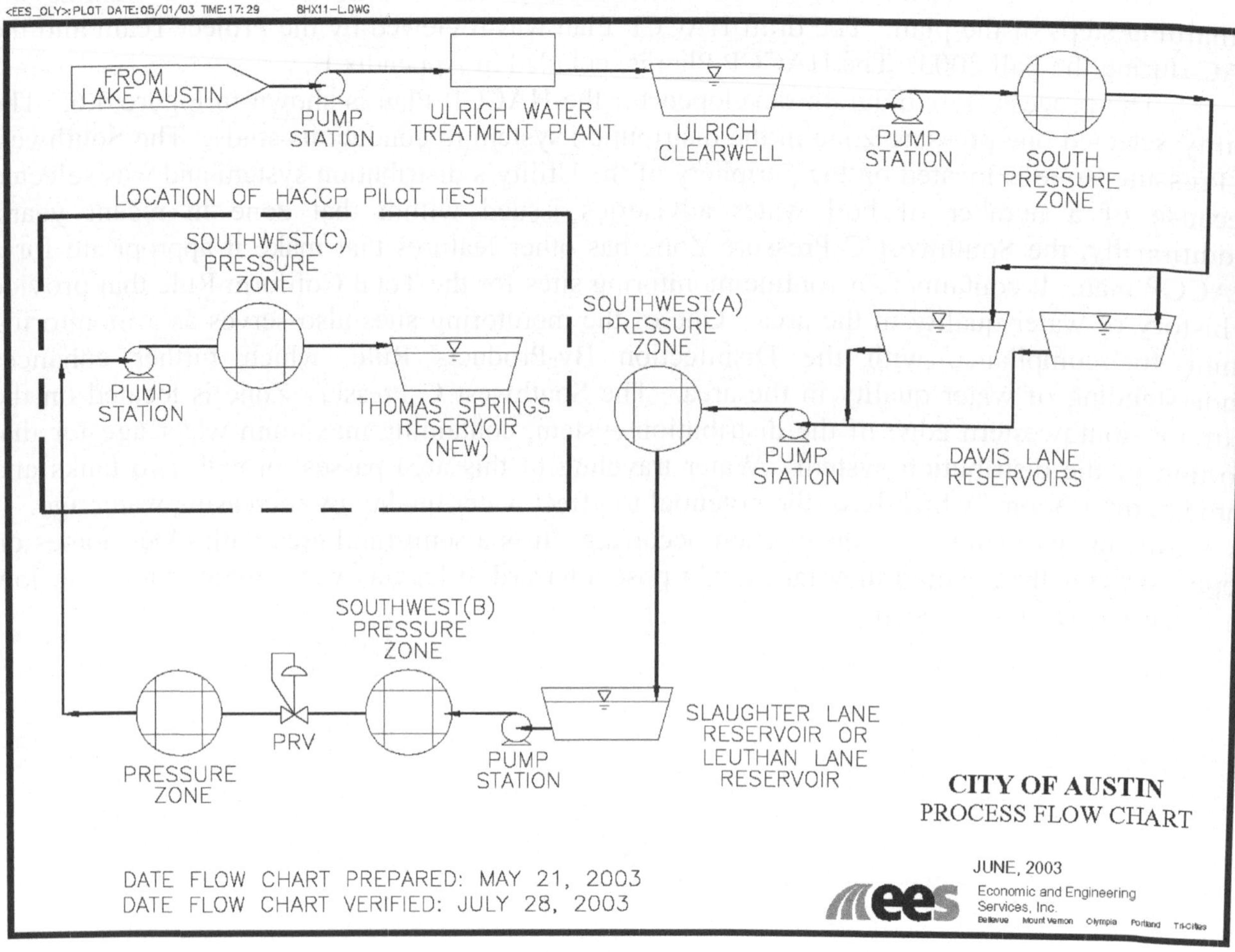

Figure 4.6 City of Austin process flow diagram

The hazard analysis (HACCP Step 6) completed at the workshop identified a number of potential hazards for the pilot study area:

- Backflow through an unprotected cross connection (risk score 25);
- Contamination at new construction sites (risk score 20);
- Backflow from failing septic systems into distribution main (risk score 9);
- Contamination due to water main break or repair (risk score 12);
- Pathogen intrusion to distribution main due to leaky sewage main (risk score 10);
- Intentional contamination of distribution system by vandals (risk score 9);
- Contamination due to main break outside pressure zone (risk score 8);
- Contamination due to pressure transient (risk score 4);
- Presence of bacteria due to low disinfectant residual (risk score 4); and
- Presence of bacteria due to zero disinfectant residual (risk score 4).

The HACCP team decided to base the pilot study on two particular hazard events that scored high marks in the hazard analysis – backflow from unprotected cross-connections and hazards at new construction sites that can potentially degrade the finished water quality. The concern for possible contamination at new construction sites focused on contractors' unauthorized operation of system valves and the effectiveness of disinfection of new mains prior to being placed into service.

Implementation of HACCP Plan

Implementation of the HACCP plan began on October 1, 2003 and ended on September 30, 2004, coinciding with the utility's 2003-04 fiscal year.

Monitoring strategies associated with a HACCP Plan do not rely solely on water quality testing but may also include inspections and checks on various databases or other utility records. For example, one hazard selected by Austin – backflow from unprotected cross-connections includes the following monitoring strategies to evaluate the effectiveness of control measures:

- Plumbing inspections and water protection surveys to identify any unprotected cross connections and to check for working backflow prevention devices;
- Plumbing inspections and water protection surveys to check on repairs of failed backflow prevention devices;
- Pressure monitoring throughout the distribution system to evaluate whether system pressure is being maintained;
- Reviews of the utility's database to check if annual inspections of backflow prevention devices were conducted; and
- Visual inspection of on-site septic systems to check for system failures.

Monitoring strategies for Austin's new construction site hazard include:

- Phone and radio contact with site personnel to check on unauthorized use of system valves;
- Visual inspection of contractor disinfection practices on new water lines;
- Water quality samples to check coliform bacteria counts and free chlorine levels of water in new mains prior to being placed into service;
- Monitoring tank water level and pressure point alarm levels that would indicate if any valves were opened between two pressure zones;
- Site inspections to check if contractors utilized the One-Call system (a communication system that provides a toll-free number for contractors/designers to call facility owners prior to excavating to prevent damage to underground facilities and utility lines) to mark water utility lines; and
- Reviews of sign-in sheets for training workshops to check whether site inspectors attended required training sessions.

Results of Pilot Study

The City of Austin's pilot study report is provided in Appendix C.

Austin found that HACCP is more complex than initially envisioned. Originally, the Utility thought that HACCP would involve identifying critical flow paths within the distribution system and monitoring these flow paths more intensively to assure water quality to downstream sites. Instead, by nature of the selected hazards, the measures used to control these hazards focused on operations and maintenance activities rather than water quality monitoring. This approach added layers of complexity to the existing monitoring program.

Austin determined that existing databases would need to be modified to facilitate the implementation of HACCP system-wide. Databases that track plumbing inspections, cross-connection inspections, waterline disinfection, and etc., were not set up with HACCP in mind. Most existing utility databases can recall data by street and City grid, but not by pressure zone. Therefore, data retrieval was undertaken manually and was cumbersome.

Austin found that implementing the HACCP plan was a lengthy time commitment including HACCP team meetings, coordination meetings with other staff, and time spent on database management. It is difficult to estimate the cost of implementing HACCP for Austin because they only tracked HACCP team meeting time and not other staff time spent implementing HACCP procedures. Because Austin's HACCP Plan only represented one small portion of their distribution system, the resource requirements for implementing HACCP system-wide were not developed.

The HACCP pilot study helped Austin to raise employee awareness on several issues:

- The need to respond quickly to main breaks in small pressure zones.
- The location of pressure zone boundaries.
- The possibility of pressure transients causing low or negative pressure in the distribution system.
- The need to maintain positive pressure at all times.
- Improved understanding of existing data sources and data management system capabilities.
- Improved understanding of distribution system issues.

No significant water quality incidents occurred in the Southwest C pressure zone during the pilot study period or the preceding 12 months so the City was unable to assess whether the HACCP Plan helped to reduce the number of water quality incidents. Although reduced numbers of customer complaints were recorded during the pilot-study period as compared to the previous 12 months, these reductions are not necessarily a result of the HACCP Plan. The HACCP training workshop did help the City complete a systematic identification of risks and control.

Sydney Water Corporation

Sydney Water Corporation supplies 20 percent of Australia's population, about four million people. It used the Woronora Water Supply System for the purpose of this study. This system has the capacity to serve a large, urban community of 200,000 people. However, during the major part of the year, it supplies only 100,000 people due to its requirements to maximize river flow within its catchment, limiting its capacity to supply the total system. This creates operational issues in relation to variable draw down from the water filtration plant and less demand within the distribution system. The system receives surface water from a fully protected, forested watershed and is treated using direct filtration and chloramination. The system was recently successfully converted to chloramination.

Training Workshop

During August 2003, Sydney Water held a two-day HACCP workshop. Attendees included planning and operational staff dealing with both treatment and distribution system operations. In addition, contract staff members were in attendance. The workshop agenda is shown in Table 4.1. In this case, workshop discussions were held as one large group rather than small breakout groups. Workshop attendees were able to complete most of HACCP steps 1 through 12 but only in an outline form. The purpose of the workshop was training rather than plan completion.

Formation of HACCP Team

At the workshop, Sydney Water decided to structure their HACCP plan along similar lines to South East Water, Victoria, Australia, by developing a series of subplans. Team members are shown in Table 4.5.

Table 4.5
Sydney Water's HACCP team

Name	Position	Role and responsibility
Corinna Doolan	Project Manager	HACCP Champion & Project Manager
Lindsay Mullard	Area Team Leader	Water Distribution System Operational Team Leader
Colin Storey	General Water Australia Principal Chemist	Treatment issues
Keith Ross	General Water Australia Operations Manager	Treatment issues
M. Govintharajah	Relationship Leader	Treatment issues
David Cooper	Water Network Officer	Reservoir subplan
Shariff Shockair	Network Management Officer	Pressure and backflow subplan
Tony Venturino	Water Quality	Discoloured water subplan Disinfection subplan
Carl Deininger	Water Operations	Water age and hydraulics subplan Maintenance subplan
Peter Cresta	Environment & Innovation	Water quality monitoring quality assurance On-line monitoring calibration
Philip Broad	Water Quality Leader	Project leadership
Dammika Vitanage	Area Water Quality Specialist	Project leadership

HACCP Plan Development

The HACCP Team worked together over several months with the HACCP Champion maintaining minutes and coordinating the activities of each person. A risk assessment was developed using an Excel spreadsheet.

The utility selected the whole of its Woronora system for the case study. The system has approximately 700 miles (1,126 km) of pipelines, 20 covered service reservoirs and about 78,000 service connections. The supply is continuous.

Subplans were developed for particular risk areas by operational and planning staff as designated in Table 4.5. The subplans were assembled into a single document and reviewed by the Project Team and the PAC during the Fall 2003. The process flow diagram developed for the HACCP Plan is shown in Figure 4.7.

The initial list of risks was reviewed and grouped into a shortlist of risks to develop management plans. Sydney Water proposed to focus its HACCP plan on the following issues relating to:

- Maintaining chloramine residuals;
- Optimizing system hydraulics;
- Securing finished water storage facilities;
- Managing discoloration;
- Reviewing main repairs to minimize risks for recontamination; and
- Maintaining system pressure & backflow prevention.

The team identified a number of items to be monitored at frequencies varying from continous through to weekly or annually (Table 4.6 and Table 4.7). These were part of the normal compliance and operational monitoring runs that were in place already within the system.

The Sydney Water HACCP plan is not included in Appendix B because the HACCP process was undertaken within the context of the organization's existing ISO 9001 quality management system. Instead, examples of Management Plans arising from the HACCP process are included as Appendix D.

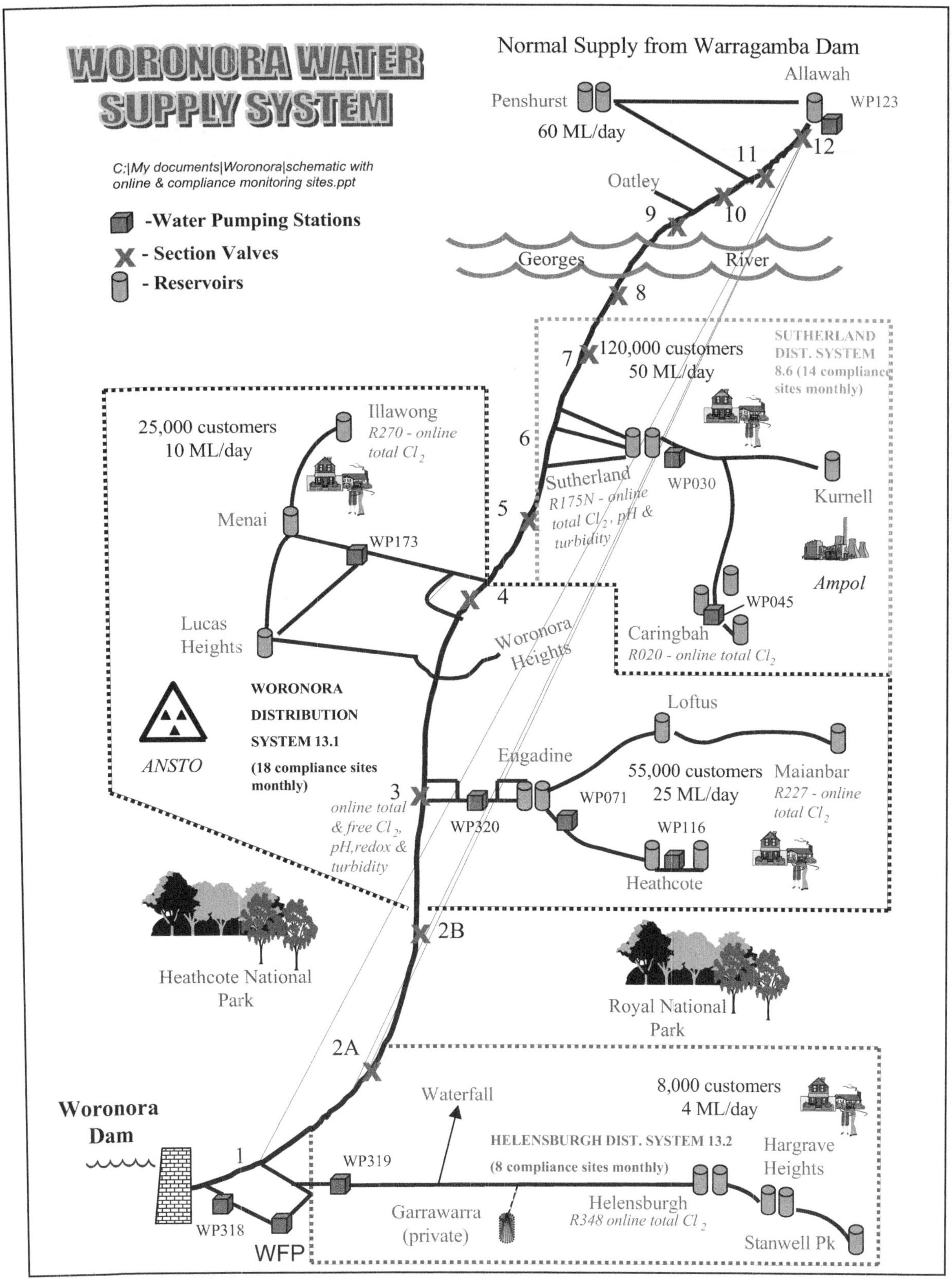

Figure 4.7 Sydney Water Woronora system flow diagram

Table 4.6
Critical monitoring parameters for Sydney Water HACCP Plan

Monitoring parameter	Criticality
Pressure	Critical
Tank level	Critical
Backflow prevention	Critical
Tank security	Critical
Chlorine:ammonia • Free chlorine • Total chlorine • Monochloramine • Dichloramine • Ammonia	Critical
Disinfection residual • Chloramine	Critical
E. coli	Critical
Nitrite	Non-critical
Nitrate	Non-critical
Total coliforms	Non-critical
Heterotrophic plate count	Non-critical

Table 4.7
Sydney Water Corporation monitoring program

Sampling location	Monitoring parameter	Monitoring frequency, type	Documentation
HWO1 - Raw Water	Chlorine	Continuous, on-line	SCADA
	E. coli	Weekly	LIMS database
	Ammonia	As above	As above
	Nitrate	As above	As above
	Nitrite	As above	As above
	Total coliforms	As above	As above
	HPC_{20}	As above	As above
HWO2 – Downstream of CWT Compliance / hand over point	As above	As above	As above
Section Valve 3	As above	As above	As above
Reservoirs	Tank level	Continuous, on-line	SCADA
• Engadine Helensburgh	Tank security	Weekly	Inspection records
• Heathcote Menai	Chlorine and speciation	As above	LIMS database
• Maianbar Illawong	*E. coli*	As above	As above
• Reservoirs R362 Inlet &	Ammonia	As above	As above
• R258 Outlet	Nitrate	As above	As above
	Nitrite	As above	As above
	Total coliform	As above	As above
	HPC_{20}	As above	As above
Bombora Ave Bundeena	Pressure	Continuous, on-line	SCADA
	Chlorine and speciation	Weekly	LIMS database
	E. coli	As above	As above
	Ammonia	As above	As above
	Nitrate	As above	As above
	Nitrite	As above	As above
	Total coliform	As above	As above
	HPC_{20}	As above	As above
Clough Ave Illawong	As above	As above	As above
Critical backflow high hazard customers	Backflow prevention unit and audit	Annually	Inspection records

Ave - Avenue
LIMS – Laboratory information management system
SCADA – Supervisory Control and Data Acquisition system

Implementation of HACCP Plan

In implementing the HACCP plan, security patrols were increased and improved chlorine residual monitoring was undertaken. In addition, more information about system operations began to be recorded on paper and management of hydraulics and water age became more systematized. The existing ISO 9001 quality system was considered to have been enhanced by the use of the HACCP system since the HACCP system helped to focus attention on the most important quality system procedures.

After the first six months of implementation, the following issues were identified as being of particular importance:

- Maintaining chloramine residuals;
- Optimizing the system hydraulics; and
- Securing finished water storage facilities.

Some issues were identified from the HACCP program monitoring. For example, a drop in chlorine residuals was identified that was related to system hydraulics. The issue was recognized as a result of the HACCP system implementation.

The spreadsheet register of risks generated as part of the HACCP plan was reviewed after six months. Upon reflection, many of the risks identified as high or medium were re-classified as low following more experience.

The monitoring results were not used to provide information on trends for several reasons. First, after many years where routine flushing was possible, the worst drought on record prevented this practice leading to an expectation of reduced water quality. Second, changes in water quality were expected to only conclusively arise following periods of several years. Therefore, in the short-term, HACCP was considered to be more of a system for monitoring to identify the need for corrective actions rather than a tool for trend analysis. The objective and subjective evaluation of the HACCP plan by Sydney Water is detailed in Table 4.8.

Sydney Water believed that the HACCP model utilized for the project was a useful tool for helping to structure water quality risk assessment and management. Sydney Water has proposed to adopt a standard operational procedure within its quality system that recommends the use of HACCP in updating its quality system.

Table 4.8
Sydney Water perspective on HACCP pilot study

Question	Answer	Basis/examples
Reduced taste/odor complaints? Reduced discolored water complaints?	No attributable change	There was a slight decrease in complaint numbers but there was also a change in the complaints recording system in October 2004. It was not possible to assess the impact of HACCP.
Reduced low pressure complaints	No change	System experiences minimal low pressure problems, usually from short-term shutdowns only.
Reduced *E. coli* or thermotolerant coliforms? Reduced total coliforms? Reduced HPCs? Reduced turbidity? Reduced low pressures?	No attributable change	During the pilot study period, many system changes were instituted as well as a change in testing methods (to Colilert). Therefore, it was not possible to determine the impact of HACCP.
Reduced water quality incidents*?	No change	No water quality incidents were experienced.

(Continued on next page)

Table 4.8 (Continued)
Sydney Water perspective on HACCP pilot study

Question	Answer	Basis/examples
Reduced chlorinator failures?	Yes	A HACCP plan for disinfection in the Woronora system was in place before the project commenced and may have helped to reduce chlorination failures.
Systematic identification and prioritization of risks and control?	Yes	Completed as part of the HACCP workshop. Following the workshop, major risks were grouped and developed into management plans.
Improved understanding of the distribution system?	Yes	The HACCP project gave the system operators a more focused direction on system operations including a more structured performance review process.
Improved understanding and maintenance of positive system pressure?	Yes	Operating reservoirs at lower levels to improve throughput without compromising system pressure.
Enhanced backflow and cross-connection prevention?	No change	Audits were already required to establish adherence to maintenance procedures.
Has monitoring become more focused?	Yes	Awareness was raised on the possible locations of risks and the required monitoring locations. Critical performance levels were determined.
Improved databases?	Yes	One data management system required changes to meet the needs of the system operators to allow storage of monthly trends.
Improved calibration schedules?	No change	Calibration schedules were already well-managed.
Better targeted reporting?	Yes	Reporting became more focused on priority areas as determined by HACCP identification of risks and critical control points.
Improved record keeping?	Yes	The system operators became more focused on reviewing the system on a regular basis and summarizing events and exceptions.
Made it easier for O&M staff to operate system?	Yes	The performance review became more structured, concentrating on critical assets.
Enhanced operator empowerment (ownership, understanding and involvement)?	Yes	Operators have been more involved in developing standard operating procedures and understanding their purpose and relevance. Operators and management have improved utilization of existing information systems.
Did additional training in water safety get introduced?	No change	No additional training was carried out.
Did existing training become less ad-hoc and more structured?	No change	Training was not part of this HACCP plan.

(Continued on next page)

Table 4.8 (Continued)
Sydney Water perspective on HACCP pilot study

Question	Answer	Basis/examples
Improved senior management involvement?	Yes	A standard operating procedure on HACCP was developed for Sydney Water as a whole.
Improved relationships between management and operators?	Partially	It is premature to assess improved relationships for the total process. However, in relation to the disinfection process, relationships have improved.
Improved clarity of accountability?	Yes	System operators have become more focused. Water quality is not seen as the exclusive domain of the water quality group.
Has communication been improved within the utility and between the utility and external stakeholders?	Yes	Communication has improved within the utility on disinfection issues for the case study area.
What are the estimated costs of implementation of the HACCP system?	$60,000	Includes internal staffing costs for meetings, site visits and system review.
Did HACCP introduce new processes and concepts?	Yes	Giving ownership to the system operators; promoting the drinking water framework.
Was good use made of existing systems to minimize duplication and maximize integration?	Yes	The existing quality assurance system was used. As a result of HACCP, it was determined that this quality assurance system needs to be streamlined.
What is the estimated value of the benefits of these changes?	Not quantified	Not applicable.
Has HACCP enhanced any environmental outcomes?	Possibly	When disinfection management improves, there is less discharge of chlorinated water to the environment.
How was the auditing (inspection, evaluation) process perceived?	Not applicable	No auditing was carried out.
Has continuous improvement been introduced?	Yes	Improvements were made to data management system and use of system performance data for decision-making. System operators have improved their focus on the daily system operations.
Has credibility and due diligence improved?	Yes	The HACCP process actively identifies system faults through ongoing review activities.

N/A – not applicable
* Water quality incident is defined as a major water quality problem leading to some kind of emergency response such as extensive dirty water events, taste and odor events or suspected contamination events.

Power and Water Corporation

Power and Water Corporation is responsible for providing energy, water, sewerage, and communications services throughout the Northern Territory of Australia. The total population of the Territory is approximately 200,000 persons. Power and Water's geographical coverage extends from the city of Darwin, in the tropical north, to the outback town of Alice Springs located in the arid climate of central Australia. Power and Water operates across a diverse range of urban centers, rural centers and remote Aboriginal communities. Darwin is the largest population center with a population of 100,000 and is supplied water primarily from a surface water reservoir in a protected catchment. The surface supply is supplemented with groundwater from a protected aquifer. The reservoir is recharged during the monsoon, wet season (November to April) and is drawn down through the dry season (May to October). The only treatment is chlorination (primary and secondary) and fluoridation. In comparison, townships in the arid region rely on a groundwater bore supply where water resources are limited. In many cases, chlorination is the only form of treatment.

The Katherine water supply system was selected for the Project pilot study. Katherine is located 186 miles (300 km) south of Darwin and has a population of approximately 10,000 persons. The utility selected the whole of its Katherine system, including the water treatment plant, for the pilot study. At the treatment plant, untreated surface water undergoes coagulation/flocculation, clarification and filtration. The filtered water is then combined with groundwater that has been aerated and the combined flow is then chlorinated and fluoridated. For part of the year during which the raw surface water turbidity is low, the plant's operation is altered and the raw surface water is only filtered. The system contains three closed storage tanks in the distribution system, approximately 61 miles (98 km) of mains and about 2,000 customer service connections.

Training Workshop

During August 2003, Power and Water Corporation held a two-day HACCP workshop. Attendees included senior management, water quality, and operational staff including those from the head office in Darwin and those operating the Katherine system. The workshop agenda was similar to the example shown in Table 4.1. Workshop discussions were held as one large group rather than small breakout groups. Workshop attendees were able to complete most of HACCP Steps 1 through 12 but only in an outline form. The purpose of the workshop was training rather than plan completion. A second workshop was also held in December 2004 to review the draft HACCP plan once written.

Formation of HACCP Team

At the workshop, Power and Water representatives agreed there would be one HACCP team with most of the work being coordinated from the Darwin head office. A smaller coordination group was also established to drive the pilot project; however, difficulties were faced with maintaining the momentum of this group. A senior manager's group was also established to provide oversight to the pilot project and implementation of HACCP across other systems.

Late in 2004 an engineer was employed and located in Katherine to facilitate implementation of the pilot project. Table 4.9 lists the team members who were involved in the pilot project at the completion of the pilot project. During the course of the pilot project, a number of the initial HACCP team members left the organization to undertake emergency aid work overseas following the tsunami in late December 2004.

Table 4.9
Power and Water Corporation's HACCP team

Name	Position	HACCP role	Responsibility in HACCP
Simon Copley	Asset Maintenance Coordinator		• Implementation of plan
Norm Cramp	Manager Water Operations	Team member Management commitment	• Review plan to recommend approval • Facilitate implementation of plan upon approval; • Verification of plan
Mark Ewin	Assistant/Water Treatment Plant Operator	Team member	• Implementation of plan
Gabrielle Halcrow	Environmental Heath Officer	Team member Technical expert	• Input to plan on area of expertise
Paul Heaton	Manager Water Facilities	Team member Management commitment	• Review plan to recommend approval • Facilitate implementation of plan upon approval
Peter Hopkins	Water Treatment Plant Operator	Team member	• Implementation of plan upon approval • Verification of plan
Noel McCarthy	Project Manager Quality Systems	Coordination Team member Technical expert	• Input to plan on area of expertise • Development of plan • Verification of plan
Leon Miles	Water Quality Engineer	Team member	• Implementation of plan • Verification of plan
Kevin O'Brien	Katherine Systems Coordinator	Team member	• Implementation of plan upon approval • Verification of plan
Declan Page	Senior Resource Planner	Team member Technical expert	• Input to plan on area of expertise • Validation of plan

HACCP Plan Development

The Project Manager of Quality Systems undertook the majority of HACCP plan development over several months. The HACCP plan is included in Appendix B. The primary process flow diagram developed for the HACCP plan is shown in Figure 4.8.

Power and Water Corporation developed both critical control points (CCPs) and non-critical control points (NCCPs). Within the distribution system the CCPs included:

- Maintenance of chlorine residuals;
- Closed storage facilities;

- Backflow prevention program; and
- Pressure maintenance practices.

The HACCP team identified a number of critical monitoring parameters to be monitored at frequencies varying from continuous to weekly as summarized in Table 4.10.

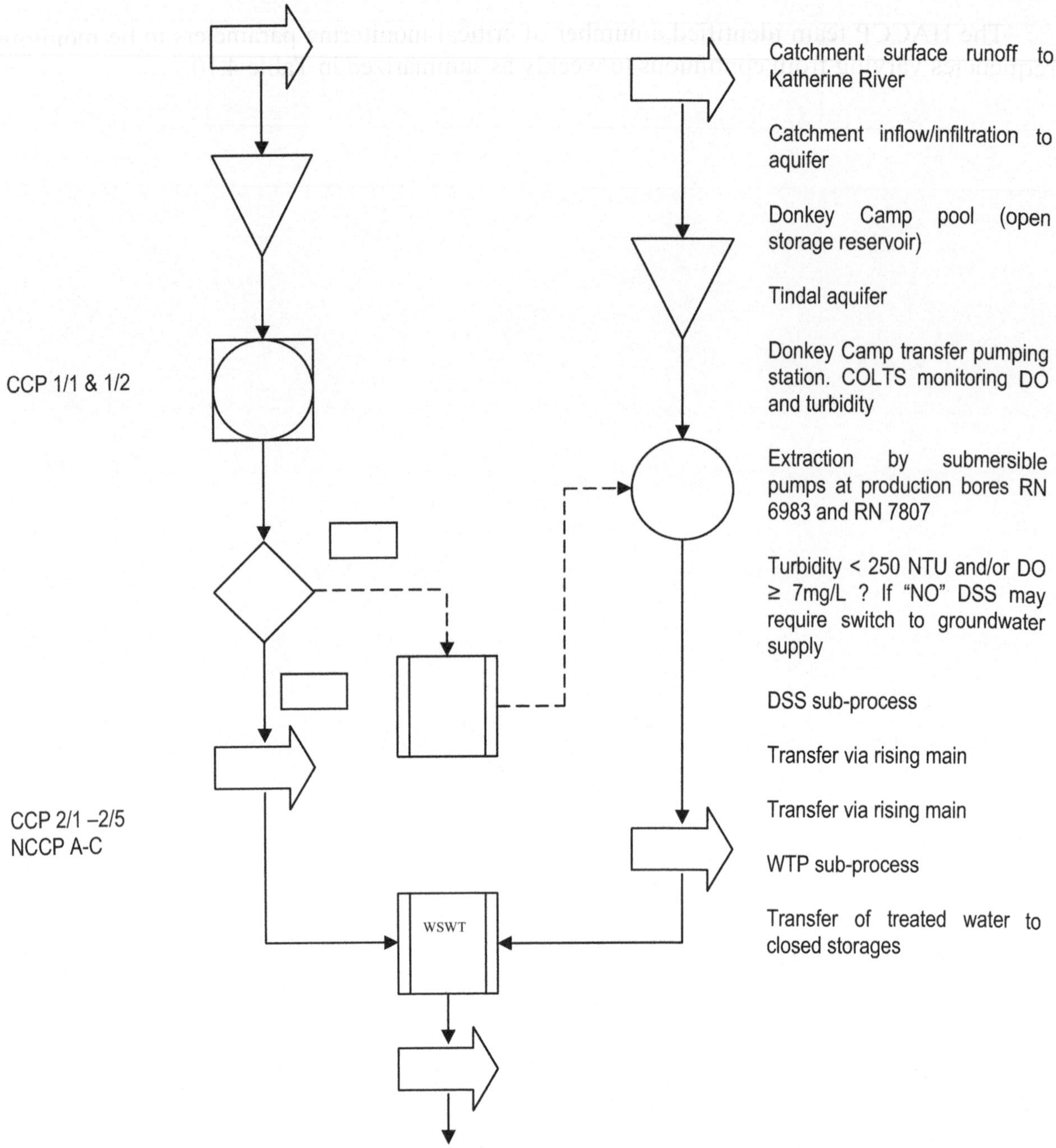

Symbol	Description	Symbol	Description	Symbol	Description	Symbol	Description
○	Operation	◇	Decision	▽	Storage	◫	Sub-process
□	Inspection	D	Delay	⇨	Transfer	🚶🚶	Product user

Figure 4.8 Power Water Corporation's Katherine supply process flow diagram

(Continued on next page)

(Continued)

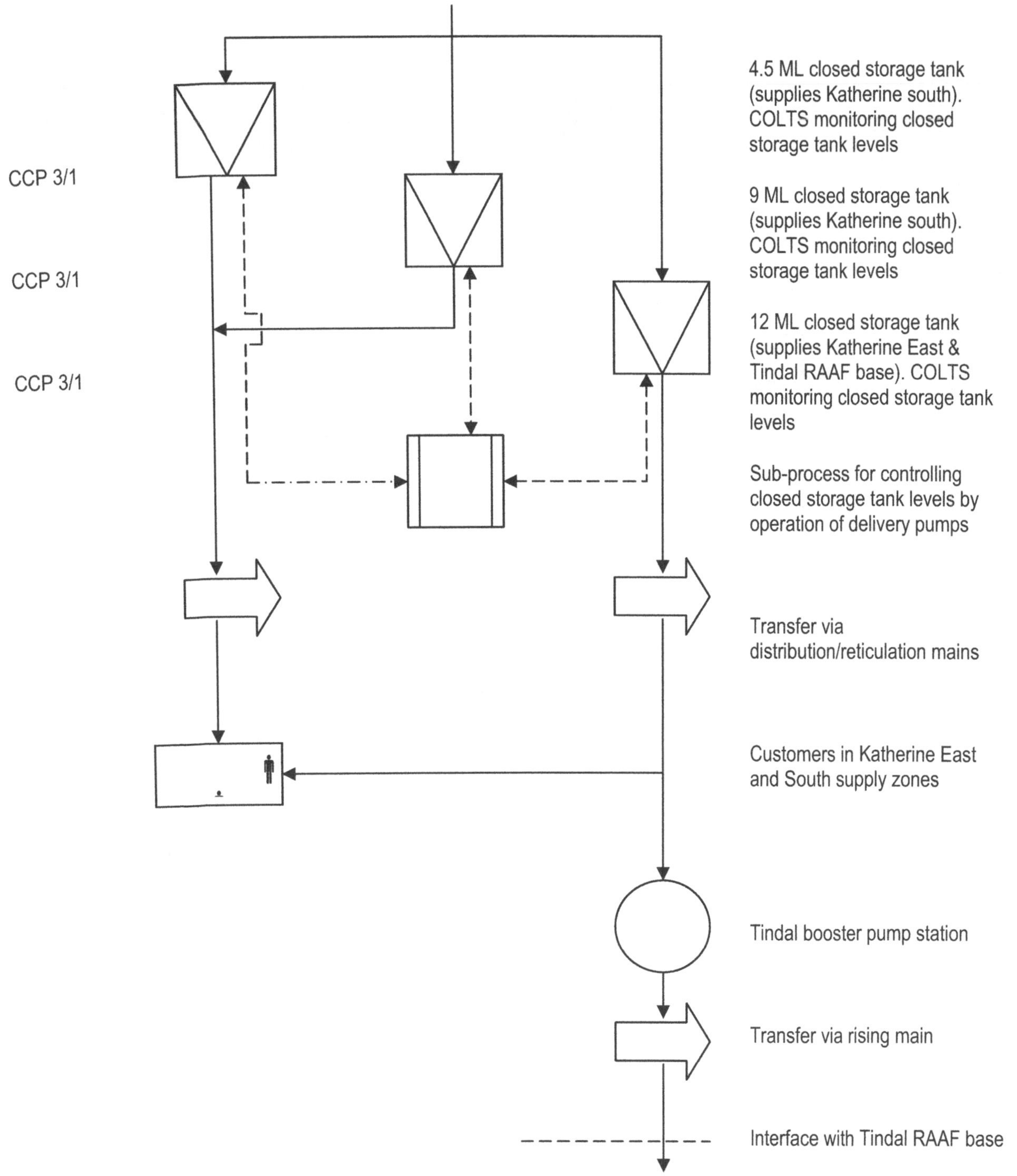

Symbol	Description	Symbol	Description	Symbol	Description	Symbol	Description
○	Operation	◇	Decision	▽	Storage	◫	Sub-process
□	Inspection	D	Delay	⇨	Transfer	🚹🚹	Product user

(Continued on next page)

Figure 4.8 Power Water Corporation's Katherine supply process flow diagram

(Continued)

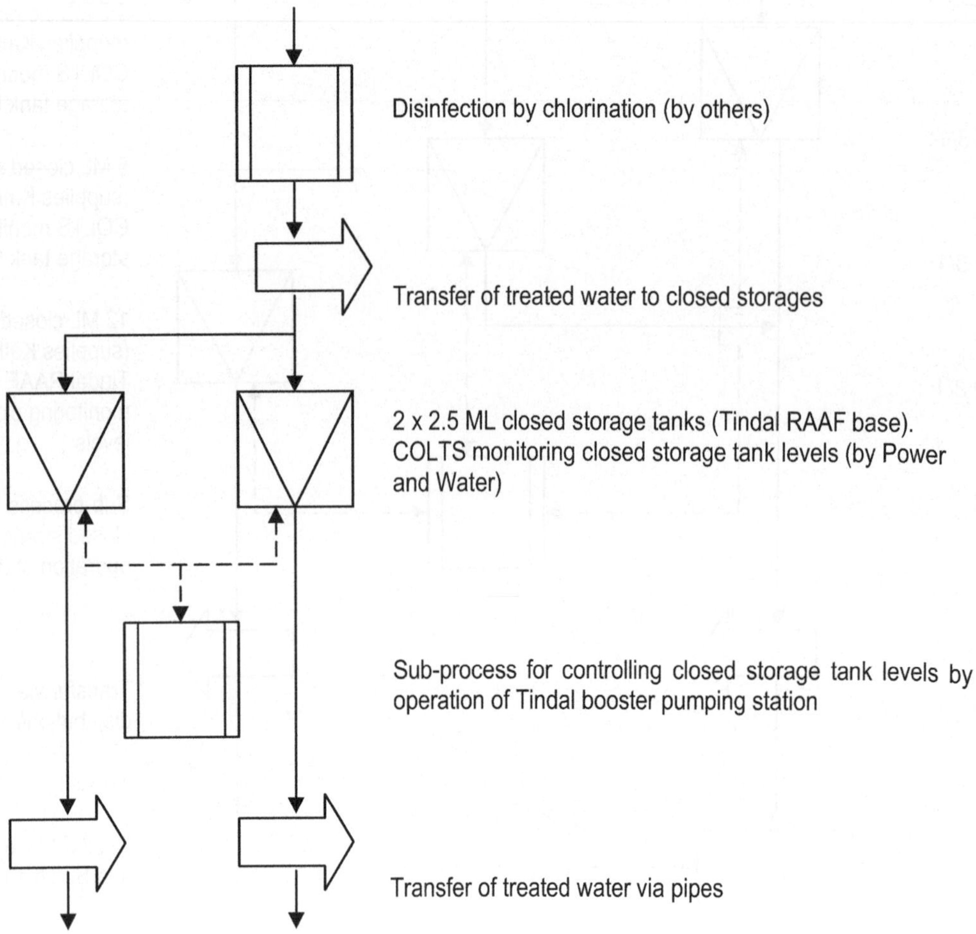

Symbol	Description	Symbol	Description	Symbol	Description	Symbol	Description
○	Operation	◇	Decision	▽	Storage	◫	Sub-process
□	Inspection	D	Delay	⇨	Transfer	🚹🚺	Product user

Figure 4.8 Power Water Corporation's Katherine supply process flow diagram

Table 4.10
Power and Water Corporation monitoring program

Sampling location	Monitoring parameter	Monitoring frequency, type	Documentation
Closed storage tanks: 1.2 MG (4.5 ML), 2.4 MG (9 ML) and 3.2 MG (12 ML)	Tank level	Continuous, on-line	SCADA reports
Tindal booster station: test point within compound	Chlorine residual	Taken week days, field measurement.	Daily water test sheets
Crawford Street: Hose tap within sewage pump station compound	Chlorine residual	Taken week days, field measurement.	Daily water test sheets
Lucy Street: test point on water service	Chlorine residual	Taken week days, field measurement.	Daily water test sheets
Tindal booster station: test point within compound Crawford Street: Hose tap within sewage pump station compound Lucy Street: test point on water service Katherine East Pump Station	Fecal coliforms or *E.coli* HPC	Weekly	Water quality database
Fire hydrant on each new main constructed by private developer	Fecal coliforms or *E.coli* HPC	Single grab sample taken and laboratory analysis performed. Connection to existing network does not proceed until pass result	Laboratory test report

SCADA – supervisory control and data acquisition

Implementation of HACCP Plan

Implementation of the HACCP plan began on April 1, 2004 and is continuing. It is Power and Water's intention to implement HACCP for the Katherine system plus seventeen other systems across the Northern Territory.

Power and Water found that the simple diagram that describes the 12 steps of HACCP belies the challenge of actually implementing HACCP properly to achieve the maximum benefits. For example, it is possible that completing steps 9 through 12 will require new processes to be established and, in most cases, documented to ensure their permanence. Power and Water endorsed the need to accept at least some shortcomings in existing operational monitoring, practices, records management and etc., and define a program of improvements (captured in the HACCP plan) to address these shortcomings, rather than try to achieve perfection with the initial HACCP plan. Of course, this approach only works if the identified shortcomings do not seriously compromise the system in the short-term.

Another issue identified by Power and Water in applying HACCP is the flexibility in translating broad HACCP principles into specifics (i.e. what exactly has to be done and how can it be undertaken). Power and Water found that using the management system standard ISO 9001 was useful to provide guidance on how to develop an effective management system to help effect the process control elements of HACCP. Power and Water also benefited from reviewing HACCP plans for other water utilities.

Power and Water identified that HACCP implementation to more than one water supply system across a large area, such as the Northern Territory, is significantly more resource-intensive than the effort required to implement HACCP for a single system. The effort required in conducting workshops, developing multiple plans, managing greater amounts of data and information, coordinating participants, undertaking audits, and monitoring progress can consume large amounts of time. Senior management commitment to such a program is essential and this must translate into clear priorities and adequate resources.

An existing, sufficiently developed ISO 9001 quality management system addressing water quality management can greatly facilitate HACCP implementation because the organization will have already defined, developed and, where appropriate, documented its management and production processes. Many of the HACCP prerequisite supporting programs would be addressed through the ISO 9001 system. For example:

- Management review – existing structures and processes would already be developed to enable review of data and information including customer complaints, exceedances/ exceptions, etc.
- Internal auditing – an established auditing system (i.e. schedules, procedures, auditing tools, trained auditors etc.) would readily accommodate HACCP-specific requirements while ISO 9001 process audits would address the compliance of core management and product realization processes.
- Measuring and monitoring – appropriate monitoring of processes and drinking water would be defined and implemented (e.g. operational and compliance monitoring programs) including effective calibration of portable, laboratory, and permanent *in situ* water quality measuring and monitoring equipment including provision of appropriate records.

- Purchasing – control over procurement activities such as the supply of drinking water treatment chemicals should be in place which can impact on production processes and drinking water quality.
- Corrective and preventive actions – well established processes should be in place for responding to exceedances/exceptions including documentation of actions taken, and evaluating effectiveness of measures and processes for managing more permanent changes.

In many cases the implementation of an ISO 9001 management system will drive the organization to address the adequacy of many of its supporting information technology systems which manage data and information (e.g. SCADA, water quality database, backflow device database, land development activity database etc.). These sources of data and information can often be the source of records to review the effectiveness of HACCP control measures.

During HACCP implementation, extra on-line instrumentation was installed at the water treatment plant and additional chlorine analyzers were installed downstream of the 1.2 MG (4.5 ML) and 2.4 MG (9.0 ML) storage facilities. Local personnel were also responsible for the provision of water supply services to six other small towns in the region, the furthest of which was located 9 hours by road from Katherine. Additional on-line instrumentation was installed in a number of these locations. During the pilot project a new chlorination facility was constructed at the water treatment plant *via* a construction contract and local personnel undertook training and assisted with commissioning. In combination these were significant commitments to undertake during the course of the pilot project.

In order to effectively apply HACCP requirements, new equipment must be calibrated, maintained and integrated with existing SCADA. Additionally, data capture processes must be established. There were challenges with ensuring this level of rigor due to Katherine being a remote supply. Small, remote towns are not likely to be a base for technicians that undertake specialized equipment calibration and maintenance services. This is an important finding from the project since HACCP often drives increased on-line monitoring of critical control point parameters. Because there is a need to support equipment maintenance and calibration as part of the HACCP supporting programs, the choice of monitoring equipment needs to be made while considering the available technical support.

In preparing the HACCP system for certification, significant additional process mapping and documentation was required at an operational level but this was incomplete by the time of the AwwaRF project completion. Certification had not been applied for or awarded at the time of writing. Power and Water was intending to implement ISO 9001 by July 2006 and this was assisting with the development of any required documentation and record-keeping for HACCP. There was a view that HACCP certification would be valuable to provide a driver for maintaining effort and focus on water quality management. The view was that if the emphasis on the HACCP system was taken away, people and processes may lose their focus and/or revert to previous practices.

In terms of performing the hazard (i.e. risk) analysis, a revised, novel, qualitative/semi-quantitative scoring matrix will be introduced in 2005/2006 for Power and Water. The novel matrix was considered by Power and Water to overcome a number of limitations inherent in the scoring matrix used by Power and Water for the pilot project. Figures 4.9 and 4.10 compare the two scoring matrices and indicate their important features.

An objective and subjective evaluation of the HACCP system by Power and Water Corporation is detailed in Table 4.11.

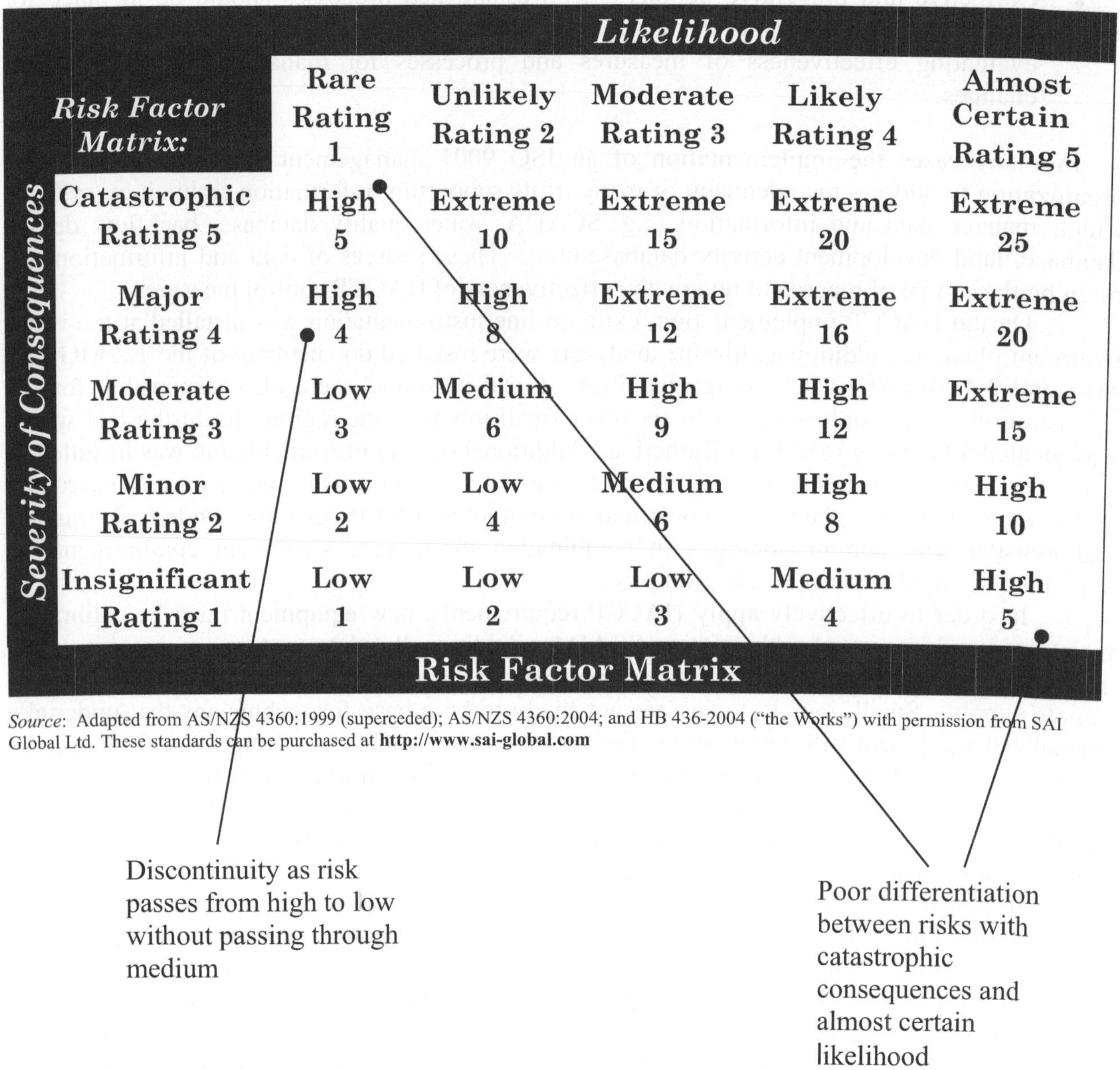

Risk Factor Matrix:	Likelihood: Rare Rating 1	Unlikely Rating 2	Moderate Rating 3	Likely Rating 4	Almost Certain Rating 5
Severity of Consequences: Catastrophic Rating 5	High 5	Extreme 10	Extreme 15	Extreme 20	Extreme 25
Major Rating 4	High 4	High 8	Extreme 12	Extreme 16	Extreme 20
Moderate Rating 3	Low 3	Medium 6	High 9	High 12	Extreme 15
Minor Rating 2	Low 2	Low 4	Medium 6	High 8	High 10
Insignificant Rating 1	Low 1	Low 2	Low 3	Medium 4	High 5

Risk Factor Matrix

Source: Adapted from AS/NZS 4360:1999 (superceded); AS/NZS 4360:2004; and HB 436-2004 ("the Works") with permission from SAI Global Ltd. These standards can be purchased at **http://www.sai-global.com**

Figure 4.9 Risk scoring matrix used for the pilot by Power and Water

Risk Factor Matrix: Severity of Consequences	Likelihood: Rare	Unlikely	Moderate	Likely	Almost Certain
Catastrophic Rating 5	High 15	Extreme 19	Extreme 22	Extreme 24	Extreme 25
Major Rating 4	High 10	High 14	Extreme 18	Extreme 21	Extreme 23
Moderate Rating 3	Mediu m 6	Medium 9	High 13	High 17	Extreme 20
Minor Rating 2	Low 3	Low 5	Medium 8	High 12	High 16
Insignificant Rating 1	Low 1	Low 2	Low 4	Medium 7	High 11

Risk Factor Matrix

Source: Adapted from AS/NZS 4360:1999 (superseded); AS/NZS 4360:2004; and HB 436-2004 ("the Works") with permission from SAI Global Ltd. These standards can be purchased at **http://www.sai-global.com**

Discontinuity removed as risk now passes from high to low by passing through medium

All cells have unique risk score

Improved differentiation between risks with catastrophic consequences and almost certain likelihood

Figure 4.10 New risk scoring matrix to be used by Power and Water following their experience with the pilot

Table 4.11
Power and Water Corporation perspective on HACCP pilot study

Question	Answer	Basis/examples
Reduced taste/odor complaints?	N/A	Pilot study period was too short to evaluate the impact of HACCP on reduction in complaints.
Reduced discolored water complaints?	N/A	Pilot study period was too short to evaluate the impact of HACCP on reduction in complaints.
Reduced low pressure complaints?	N/A	Pilot study period was too short to evaluate the impact of HACCP on reduction in complaints.
E. coli or thermotolerant coliforms?	No change	Already observe a high level of compliance.
Total coliforms?	No change	Already observe a high level of compliance.
HPCs?	No change	Already observe a high level of compliance.
Turbidity?	No change	Already observe a high level of compliance.
Low pressures?	No change	The network has adequate pressure.
Reduced water quality incidents?*	No change	No significant water quality event occurred during the pilot study period.
Reduced chlorinator failures?	N/A	Chlorinators are not used in the distribution system.
Systematic identification and prioritization of risks and control?	Yes	A risk assessment was undertaken at the workshop.
Improved understanding of the distribution system?	Yes	Additional pressure monitoring was undertaken to assist with network modeling.
Improved understanding and maintenance of positive system pressure?	Yes	All team members are more aware of the implications of system pressure loss.
Enhanced backflow and cross-connection prevention?	Yes	HACCP can drive the setting of monitoring requirements & performance measures that demonstrate program effectiveness, similar to "critical limits".

(Continued on next page)

Table 4.11 (Continued)
Power and Water Corporation perspective on HACCP pilot study

Question	Answer	Basis/examples
Has monitoring become more focused?	Yes	On-line chlorine analyzers have now been installed downstream of two distribution service storage facilities. In addition, other on-line monitoring has been installed at the water treatment plant to improve process control.
Improved databases?	No change	This item was outside the scope of the pilot project. However, the project has highlighted that information technology solutions are required in a number of areas to successfully implement HACCP (particularly if it is applied to multiple systems) e.g. 1) risk management software to register risks, controls, track improvements, changes in risk etc. 2) water quality database capable of accepting operational monitoring data (e.g. on-line data, field testing), water quality monitoring data, as well as recording exceedances and tracking corrective action.
Improved calibration schedules?	No change	Some difficulties were identified in calibrating on line systems in remote areas. These difficulties have broader implications for HACCP implementation in regional areas.
Better targeted reporting?	N/A	Not assessed.
Improved record keeping?	Yes	A variety of new forms have been developed to improve record keeping; however, it is recognized that upgraded information technology systems are needed to more efficiently assist record management.
Made it easier for O&M staff to operate system?	No change	Implementation is still ongoing.
Enhanced operator empowerment (ownership, understanding and involvement)?	Yes	Operator understanding of risks was enhanced.
Did additional training in water safety get introduced?	No change	Training was not a major emphasis of the pilot project.
Did existing training become less ad-hoc and more structured?	N/A	Training was not a major emphasis of the pilot project.
Improved senior management involvement?	Yes	Two senior managers are members of the HACCP team. Five senior managers make up a higher level management group that oversees implementation of HACCP.

(Continued on next page)

Table 4.11 (Continued)
Power and Water Corporation perspective on HACCP pilot study

Question	Answer	Basis/examples
Improved relationships between management and operators?	Potentially	This was not a major emphasis of the pilot project; however, through ongoing engagement, this outcome is anticipated.
Improved clarity of accountability?	Potentially	Documentation of a HACCP plan setting roles & responsibilities of individuals has the potential to improve accountability; however, successful implementation requires that all team members have full understanding.
Has communication been improved within the utility and between the utility and external stakeholders?	Yes	Communication improvement within the utility has been greater than between the utility and external stakeholders. The HACCP team included an environmental health officer so there has been some limited communication of our work to their department. Also our annual report provides some information on HACCP implementation.
What are the estimated costs of implementation of the HACCP system?	Not quantified	Unable to quantify in dollar terms because implementation hours were not tracked. To successfully implement HACCP, a dedicated coordinator is required as well as a commitment/initiative from all team members to undertake their part. We initially had two workshops (one training & one workshop to agree on a final HACCP plan) which involved the team and others. This required organizing a dozen or so people to meet in Katherine which is a three hour drive from Darwin. Where there is more than one water supply system and systems that are widely dispersed geographically, then implementation of HACCP is far more complex. In addition, substantial time has been invested coordinating and attempting implementation. Also, various team members have been working on specific improvements (e.g. installing new monitoring equipment, strengthening catchment management). We initially started with a monthly coordination meeting (not on-site), a senior managers meeting every 6-7 weeks, plus about 5 site visits. It proved difficult to convene the team during implementation. All this was difficult to maintain but is required to initiate implementation. On-site we have found implementation difficult without a full time presence and have employed another engineer to drive implementation. Information technology solutions are necessary.
Did HACCP introduce new processes and concepts?	Yes	Internal auditing (focus on documentation and record keeping); Semi-quantitative risk assessment; Increased awareness of pressure transients leading to intrusion.
Was good use made of existing systems to minimize duplication and maximize integration?	No	HACCP highlighted deficiencies in many existing areas. Applying HACCP is not as straightforward as working through the 12 steps. For successful implementation, many processes and systems need to be established (e.g. equipment calibration, complaint handling, water quality exception reporting, data management).

(Continued on next page)

Table 4.11 **(Continued)**

Power and Water Corporation perspective on HACCP pilot study

Question	Answer	Basis/examples
What is the estimated value of the benefits of these changes?	Not quantified	Our expectation is that operational costs will rise as a consequence of new monitoring equipment installed. (i.e. increased calibration requirements), change in treatment process requiring increased chemical use.
Has HACCP enhanced any environmental outcomes?	Potentially	HACCP is expected to result in improved environmental outcomes once new arrangements are finalized, agreed upon and implemented. A spin off from HACCP has been that it required various processes & activities be examined. This included review of existing new mains construction & commissioning processes which re-examined disposal of chlorinated water.
How was the auditing (inspection, evaluation) process perceived?	Too early to say	No identified problems, although only one audit completed.
Has continuous improvement been introduced?	Too early to say	Continuous improvement would require far more time to establish and would only begin to emerge with more frequent auditing, plan reviews etc.

* Major water quality problems leading to some kind of emergency response such as extensive dirty water events, taste and odor events, or suspected contamination events.
N/A – not applicable
O&M – operations and maintenance
HPCs – heterotrophic plate count bacteria

CONCLUSIONS

The pilot utilities found that the implementation of HACCP to water supply distribution was feasible and practical, but that the time and resource requirements were greater than originally anticipated.

The resource constraints were most acutely noted by the experience of the South Berwick Water District. For this very small utility, with only three full time staff members working during the latter part of the pilot, developing and implementing a full HACCP system was not achievable given the many day-to-day duties and activities of the utility staff. Given the genuine commitment of utility staff to the HACCP pilot, it is likely that this is a fundamental difficulty in the implementation of HACCP within small water utilities. It is important to note that these difficulties are likely to be common to the implementation of any systematic process control system, whether it be named HACCP or something else, so the solutions to these problems are likely to be more generally relevant. If HACCP, or any other systematic process control system, is to be promoted for small water utilities, there are several changes that could be made to enable implementation to be successful:

- Increase the number of employees on staff within small utilities to provide a critical mass of personnel.
- Provide temporary, contracted, support for a sufficient time period to implement the system, funded by a third party, if required.
- Provide generic HACCP plans and guidance combined with very explicit guidance, support and tools to help utilities implement the systems in practice.
- Provide support from the state or regional regulatory and support organizations.

For the larger utilities, there were sufficient resources to implement HACCP although there was a requirement to have some specific resources reasonably dedicated and assigned to the HACCP implementation task. Importantly, the preparation of the HACCP plan *per se* was not resource-intensive. Rather, it was the implementation of the HACCP plan to create a functional HACCP system in practice that required the resources. For example, the HACCP plan may recognize the need for calibration of monitoring equipment – a relatively simple item to capture within the plan. However, implementing adequate calibration processes in practice can be a significant task. Therefore, it can be concluded that implementing HACCP, or any systematic water quality process control system of similar rigor, is likely to be a significant undertaking for a water utility. Further, it is the implementation rather than the planning that is the resource intensive component of the overall HACCP process.

Despite the greater than anticipated resource requirements for implementation, the participating utilities concluded that the HACCP process is valuable. The development of the HACCP plan was useful in honing in on the most important risks and process controls for water quality management. The implementation of the HACCP system was useful in driving improvements in water quality management. The activities required to implement a HACCP plan are all sensible ones that would contribute to increased water quality security. Therefore, the decision about whether or not to implement HACCP is based on an assessment of benefits versus costs.

CHAPTER 5
CASE STUDIES

INTRODUCTION

Five of the leading water utility participants of the Cooperative Research Centre for Water Quality and Treatment (CRCWQT) in Australia have independently developed and certified HACCP plans for their water distribution systems. These utilities include:

- Brisbane Water (in the state of Queensland);
- City West Water (in the state of Victoria);
- Gold Coast Water (Queensland);
- South East Water (Victoria); and
- Yarra Valley Water (Victoria).

Whereas the four "Pilot Study Utilities" participating utilities for this project developed their HACCP plans during the project, the five "Case Study Utilities" described in this chapter completed their HACCP plans up to seven years before the report was written. The purpose of this chapter is to provide additional guidance based on this accumulated experience with the application of HACCP to water distribution.

All five case study utilities provide drinking water to major Australian cities as illustrated in Figure 5.1. Brisbane Water, serving Brisbane, and Gold Coast Water, serving Gold Coast, are located in a sub-tropical climate where as South East Water, Yarra Valley Water and City West Water serve different parts of Melbourne and its surrounding communities which have a temperate climate.

Brisbane Water and Gold Coast Water are part of the local government authority and utilize water from unprotected or partially protected watersheds. The three utilities serving the Melbourne area are corporations owned by state government and purchase water from a wholesaler, Melbourne Water Corporation. Additional information on the nature of the five case study utilities is provided in Table 5.1.

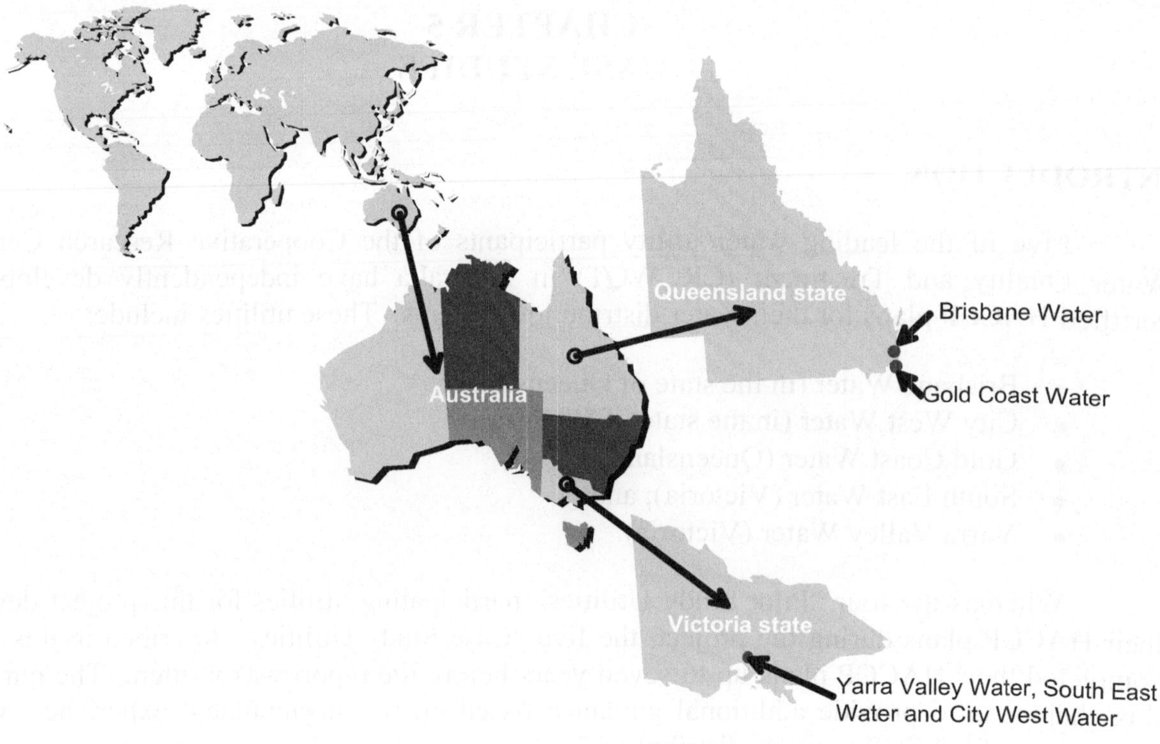

Figure 5.1 Map illustrating the location of the five case study utilities

Table 5.1
Case study utilities - system descriptions

Criterion	Brisbane Water	City West Water	Gold Coast Water	South East Water	Yarra Valley Water
Population (connections)	1.0 million + (0.4 million +)	0.6 million +	0.5 million	1.3 million (0.6 million)	1.6 million (0.63 million).
Ownership	Local government (city council) business unit	State-owned corporation (Victoria, Australia)	Local government (city council) business unit	State-owned corporation (Victoria, Australia)	State-owned corporation (Victoria, Australia)
Origin of source water	Surface water dam storages, unprotected catchments	From the wholesaler, Melbourne Water. Protected catchments (70%); unprotected catchments (30%)	From Hinze Dam system (85%) semi-protected catchment. 15% from Brisbane Water (unprotected catchment)	From the wholesaler, Melbourne Water. Protected, forested catchments	From the wholesaler, Melbourne Water. Protected catchments (75%); unprotected catchments (25%)
Treatment	100% flocculation, filtration, pH correction, chloramination	70%: chlorination, pH correction, fluoridation; 30%: filtration plus as above	2 water treatment plants with coagulation, filtration, and disinfection Brisbane Water contribution is similarly treated	Melbourne Water, chlorination, pH correction, fluoridation	Melbourne Water: Chlorination (28 zones) Chloramination (3 zones) Ultraviolet disinfection (1 zone) Microfiltration (2 zones) pH correction and fluoridation Approximately 25% conventionally filtered
Distribution system management	In house.	Planning in house. Field operations via maintenance contractor	In house	Planning in house. Field operations via maintenance contractor	Planning in house Field operations via maintenance contractor
Water quality monitoring	In house.	In-house management. Sampling and testing by laboratory contractor	In house lab for common chemistry and microbiology. Specialized tests outsourced	In-house management. Sampling and testing by laboratory contractor	In-house management. Sampling and testing by laboratory contractor
Internet address	brisbane.qld.gov.au	citywestwater.com.au	goldcoastcity.com.au	southeastwater.com.au	yvw.com.au

HACCP IMPLEMENTATION

The five participating utilities provided information on the implementation process as summarized in Table 5.2. Implementation issues are discussed in the following section.

Implementation of Other Management Systems Prior to HACCP

Most utilities that have gained HACCP certification have done so after some core management systems had been developed and implemented. These management systems helped the utility to gain management control of people and processes. The need for management systems and other supporting programs was described in Chapter 3.

International Organization for Standardization (ISO) Management Systems

There has been great pressure from customers, regulators and the broader community for Australian water utilities to transparently and objectively demonstrate good practices with respect to environmental protection, worker safety and water quality.

The ISO has developed a number of generic management systems tailored to particular aspects of organizational management activity. For example:

- ISO 9001 covers "quality management," that is, the quality of products or services produced by an organization. Water utilities have been using ISO 9001 systems to help drive improvements in all aspects of customer service and operations, including water quality.
- ISO 14001 covers "environmental management," that is, the impact that an organization has on the environment. Water utilities have been using ISO 14001 certification to drive and demonstrate good environmental protection in general, and in particular, for their sewage management activities.
- ISO 4801 (and the various state and Australian equivalents) covers "occupational health and safety." Because water utility staff sometime undertakes potentially dangerous work, utilities have been using ISO 4801 and equivalent systems to help drive and demonstrate sound occupational health and safety practices.

Table 5.2
HACCP implementation and maintenance

Criterion	Brisbane Water	City West Water	Gold Coast Water	South East Water	Yarra Valley Water
Year certified	2000 (treatment) 2003 (distribution)	2000 (distribution)	2001 (treatment and distribution)	1999 (distribution)	1999 (distribution)
Scope of HACCP Plan	Wholesaler interface to customers' meters	Wholesaler interface to customers' meters	Full system from dams to customers' meters	Wholesaler interface to customers' meters	Wholesaler interface to customers' meters
Other management systems in place prior to HACCP	ISO 9002[1] Quality System; ISO 14001 Environmental System; AS ISO 4801 Occupational Health & Safety NATA Laboratory Accreditation	ISO 9002* Quality System; ISO 14001 Environmental System; AS ISO 4801 Occupational Health & Safety	ISO 9001 Quality System; ISO 14001 Environmental Management System	ISO 9001 Quality Management System (since 1995) 14001; Environmental Management System (since 1996)	ISO 9001 Quality Management System (since 1995). 14001 Environmental Management System (since 1996)
Management systems adopted after HACCP	Integrated Management System	None	Member of Water Treatment Alliance (Australian version of Partnership for Safe Water)	Integrated Management System to streamline and integrate all systems (2002); State Occupational Health and Safety Management System (since 2000)	AS ISO 4801-Occupational Health and Safety (2003); AS7799 -Information Security Management System (2004)
In-house resources used to implement HACCP	Scientific Services Manager plus HACCP co-ordinator	Water Quality Scientist	Director appointed a HACCP champion to create system in consultation with operational staff	Mostly done in house. Chief Executive Officer and General Manager support was established. Operations (Field Services) Division appointed a HACCP Champion who convened the high-level in-house HACCP team and sat on all five in-house sub-teams (distribution, disinfection, backflow, reservoirs, new extensions)	Management commitment from Managing Director and General Manager; HACCP Coordinator responsible for overall management of HACCP; Network Operations (CCP management) and Customer operations (backflow prevention) divisions undertake relevant individual tasks

(Continued on next page)

Table 5.2 **(Continued)**
HACCP implementation and maintenance

Criterion	Brisbane Water	City West Water	Gold Coast Water	South East Water	Yarra Valley Water
Duration of development and implementation of HACCP	Initially staged for treatment and then distribution. 12 months initially + 12 months	One month.	12 months to prepare plans and procedures for dams, two water plants and distribution system	Three months	Three months
Internal resource requirement for plan development and implementation	HACCP coordinator 20% 24 months, manager 5% 24 months, various team members	0.75 FTE for one month	HACCP champion full time; Approximately 10 team members contributed occasional verbal advice upon request.	HACCP champion: full time; team members (20 in total), 1/5 time; Total 5 FTE for 3 months	HACCP coordinator: full time - 3 months; team leader – full time -1 month; team members (10) – 3 days for workshops.
Ongoing internal resource requirement for maintenance of HACCP plan	0.5 FTE	0.2 FTE	HACCP Champion full time	HACCP Champion: 1/3 time Team members (20 in total), 1/10 time Total 2.33 FTE ongoing.	HACCP Coordinator (15%) team members (10) – (5 %)
Contracted support	Pre-audit and audit by independent firm	Pre-audit and audit by independent firm	Pre-audit and audit by independent firm	Coaching-style support from outside consultants. Certification pre-audit and audit from independent firm	External consultant was engaged to undertake a detailed risk assessment prior to HACCP Plan development External consultant with HACCP expertise in the food industry facilitated the plan development.
External costs during plan development and implementation	Funded through recurrent expenses	$13,000 (Australian Dollars) ($10,500 US Dollars)	$15,000 (Australian Dollars) ($12,000 US Dollars) in instruments and $90,000 (Australian Dollars) ($72,000 US Dollars) per year increase in testing budget. (61 out of 74 urban tanks were never being tested prior to	$35,000 (Australian Dollars ($28,000 US Dollars) (training and coaching costs)	Total cost approximately $40,000 (Australian Dollars) ($31,000 US Dollars)

(Continued on next page)

Table 5.2 (Continued)
HACCP implementation and maintenance

Criterion	Brisbane Water	City West Water	Gold Coast Water	South East Water	Yarra Valley Water
			certification.)		
External costs during plan maintenance	Audit fees	$3,000 (Australian Dollars) ($2,500 US Dollars) per year (audit fee)	Approximately $6,000 (Australian Dollars) ($5,000 US Dollars) per year in external audit costs	$3,000 (Australian Dollars) ($2,500 US Dollars) per year (audit fees)	Audit fees $3,000 (Australian Dollars) ($2,500 US Dollars) per year. Annual staff training using an accredited HACCP trainer $1,500 (Australian Dollars) ($1,200 US Dollars)
Employee Training	NATA HACCP Course, DPI Centre for Food Technology used to train staff	No formal training provided	Operational and management staff was given initial awareness training. HACCP champion was an experienced quality auditor	Around 25 staff and selected stakeholders (regulator, lab, contractor, bulk water supplier) were trained up front via a tailored 2-day HACCP for water course HACCP champion trained in quality auditing	About 20 staff and maintenance contractor personnel trained annually HACCP coordinator trained on development, validation, verification and auditing of HACCP using a licensed training provider

* ISO 9002 is a former ISO Quality System that has been phased out. ISO 9002 applied to "production" processes whereas ISO 9001 applies to "design and production."
CCP – Critical Control Point
FTE – Full-Time Equivalent
ISO – International Organization for Standardization
NATA – National Association of Testing Authorities, Australia

At least two of the above systems had been implemented prior to the implementation of HACCP systems by case study water utilities. The existence of these management systems within water utilities greatly facilitated the implementation of HACCP:

- Cultural change: the organizational culture had moved to a point where utility staff was used to operating within a management system environment. Utility staff was familiar with concepts such as accountability, external and internal auditing, the use of standard operating procedures, exceedance reporting, training and competency standards, calibration, quality assurance, etc.
- Documentation and record-keeping: the additional requirements of HACCP with respect to documentation and record-keeping are often quite small once an organization has experience with a management system. For example, South East Water only required six additional procedures to be documented in implementing its HACCP plan.
- Cost – The cost of implementing HACCP may be lower for utilities that have previously implemented other quality management systems.

Integrated Management System

In practice, the case study utilities did not operate multiple, separate systems for quality management, occupational safety, water quality and safety, and environmental considerations. These separate systems were captured, as required, within an integrated management system (IMS). The principal benefit of an IMS, as identified by water utilities, was the avoidance of duplication, leading to reduced staff time and costs, and improved process integration. For example, if a utility has an SOP for a particular process, that same SOP is developed with consideration being given to the implications for staff safety, environmental impacts and drinking water quality and safety.

HACCP is one of the four common management systems that are captured within an IMS and, therefore, is readily integrated and audited as part of an IMS. Integrated audits generally took place whereby utilities could seek certification for ISO 9001, ISO 14001, AS/NZS 4801 (an occupational health and safety standard), and HACCP, as required. This integration avoided the need for a whole extra audit process. However, one or more additional auditors was usually added to the IMS auditing team to ensure sufficient expertise was available in HACCP, water safety and quality issues.

Some utilities, such as Melbourne Water (not one of the case study utilities), no longer describes their HACCP plans as "HACCP" plans. Instead, a "Water Quality Management Plan" was developed and certified against both HACCP and ISO 9001. Similarly, City West Water obtained its certification for "HACCP-9000" which represents an integrated HACCP and ISO 9001 quality and safety management system. Terms such as 'Water Safety Plan" (Davison and Bartram 2004) are also used.

Costs of Implementing a HACCP System

The reported cash costs of implementation of HACCP systems varied between utilities:

- One-time cash costs for HACCP training, coaching, technical advice and documentation varied from $0 (Brisbane Water) to $70,000 (Australian Dollars)

($55,000 US Dollars) (Yarra Valley Water). The variability is primarily explained by the difference in the extent to which utilities used external service providers to undertake some of the tasks.

- Recurrent cash costs associated with increased water quality monitoring and instrumentation varied from $0 (Brisbane Water) through to $105,000 (Australian Dollars) ($84,000 US Dollars) (Gold Coast Water) on an annual basis. The variability is related to the extent to which additional monitoring or operational costs arose from the increased risk management activities that the HACCP team deemed appropriate.
- Some utilities, such as Yarra Valley Water, adopted the recommended practice of annual HACCP awareness training, requiring an additional recurrent cash cost of $1,500 (Australian Dollars) ($1,200 US Dollars) annually. New staff, or staff that are losing familiarity with HACCP, attend the training course each year.

Non-cash costs of HACCP implementation varied between utilities. Implementation almost always required a full time, dedicated officer, for at least one month and up to twelve months. For example, City West Water required 0.75 a full-time equivalent (FTE) for one month, while Gold Coast Water utilized 1 FTE for twelve months. The variability is related to the extent to which tasks were split between a core coordinator and other staff, as well as the amount of work required to implement the HACCP plan. For example, City West Water had completed the vast majority of tasks required by HACCP before the organization formally began to undertake its HACCP process and was able to complete the HACCP plan within one month.

It is important to note that these utilities had all implemented other quality management systems prior to implementing HACCP and therefore, their costs may not be comparable to a utility implementing only HACCP.

MAINTENANCE OF HACCP SYSTEMS

Active maintenance of the HACCP system and its good water quality management practices is primarily required to prevent complacency leading to the breakdown of the controls. In addition, HACCP systems are intended to respond to a variety of inputs by facilitating the implementation of ongoing improvements.

In terms of activity, maintenance of the HACCP system involves internal auditing and other forms of verification, issue identification, and the implementation of improvements. In addition, adaptation of the HACCP system is required in response to water supply system changes as well as in response to important new knowledge or information about water quality. Once certified, a utility can expect to receive external certification audits at either 6-month or annual intervals in order to maintain its certification.

Case study information related to maintenance is summarized in Table 5.2. Some utilities commented that although external audits require a significant time commitment, they do provide a source of discipline as well as help in identifying opportunities to improve water quality and safety. In one case, a utility commented that its certification was almost lost at a time when other priorities led to reduced activity in the maintenance of the HACCP system and the auditing process had the benefit of alerting the organization to the issue.

The cost of maintaining a HACCP system in a medium sized water utility was estimated by the case study utilities as follows:

- Recurrent cash auditing costs were consistently between $3,000 (Australian Dollars) and $6,000 (Australian Dollars) ($2,500 to $5,000 US Dollars) per year with audits requiring from three to five days.
- In terms of internal costs, maintenance of the plan required between 0.2 FTE on-going (City West Water) up to more or less one FTE on-going (South East Water, Yarra Valley Water and Gold Coast Water).

BENEFITS OF HACCP SYSTEMS

All five case study utilities noted that HACCP implementation was used to provide a coordinating framework for the implementation of risk and quality management systems to achieve water quality improvements. The HACCP system was a means of consolidating and gaining certification for these improvements. Importantly, all five case study utilities have retained their HACCP certification since it was awarded, seeing it as a net positive.

Two of the five utilities, Brisbane Water and City West Water, did not observe significant improvements as a result of HACCP system implementation. Both utilities believed that they had already undertaken the vast majority of activities required by HACCP before starting the HACCP process itself. Despite these limitations, the process was seen as positive by both utilities. HACCP was considered a likely contributor to improved water quality. For example, City West Water noted that HACCP implementation led to increased staff training with respect to water quality and awareness.

A number of tangible benefits from the implementation of HACCP were perceived by three of the five utilities: Gold Coast Water, South East Water, and Yarra Valley Water. However, specific "improvements" were considered difficult to attribute entirely to HACCP implementation for two key reasons:

- Water quality improvement activities may have been undertaken without HACCP being a driver for change.
- Tangible water quality data that may provide useful indicators of water quality improvements are highly variable due to climate and other sources of "noise", making it very hard to conclusively detect effects and even harder to attribute them to any one possible contributory cause.

Despite the difficulties in establishing cause and effect and quantitative relationships, the case study utilities identified a number of benefits that were significantly related to HACCP implementation:

- Reductions in customer complaints;
- Improvements in water quality;
- Improvements in process performance;
- Improvements in work processes;
- Cost savings;
- Improved understanding of risks and risk management;
- Improvements in documentation and recordkeeping; and
- Demonstration of due diligence.

These benefits are discussed in the following sections. Details are provided in Tables 5.3, 5.4, and 5.5 for Gold Coast Water, South East Water, and Yarra Valley Water, respectively.

Table 5.3
Gold Coast Water Perspective on HACCP Application to Distribution System

Question	Answer	Basis/examples
Reduced taste/odor complaints?	Yes	See data listed under discolored water complaints.
Reduced discolored water complaints?	Yes	1999-2000: 2,116 water quality complaints. 2003-2004: 730 (complaints in year 2004 calculated on a greater scope and with better data capture technique).
Reduced low pressure complaints?	N/A	Not measured as part of HACCP.
Reduced *E. coli* or thermotolerant coliforms?	Yes	Prior to HACCP (prior to 2001), 82% of storage tanks were not tested. Therefore, past data are misleading but compliance with industry standards was nevertheless very high. 2002-2003: 2% positives. 2003-2004: 1.5% positives. Change from fecal coliform to *E. coli* February 2004.
Reduced total coliforms?	Yes	2002-2003 and 2003-2004 results > 95% pass. No data for 1999-2000 but testing regime was very selective and non-representative at that time.
Reduced HPCs?	Maybe	Long-term average for Gold Coast is around 50 CFU/mL with a mean of <10 CFU/mL. This has not significantly changed.
Reduced turbidity?	Yes	At water plants, turbidity has decreased since HACCP from around 0.2-0.3 NTU to now being routinely < 0.1 NTU. Data consistency has improved and the magnitude of variations has been reduced. Distribution system turbidity values do not vary much. The mean turbidity was 0.4 NTU for 2001-2002 and 2003-2004.
Reduced low pressures?	N/A	Not assessed as part of HACCP.
Reduced water quality incidents?*	Yes	Prior to HACCP, incidents were not well-captured. In the 4 years since HACCP was implemented, there have been between 4 and 8 incidents per year. However, this is subjective because the criteria for declaring an incident is subjective. An incident to some may be business as usual for others.
Reduced chlorinator failures?	N/A	Complete failures are rare (Only 2 failures in 4 years); however, deviations from tighter HACCP limits have been slightly more common. These would not have been noticed prior to HACCP.
Systematic identification and prioritization of risks and control?	Yes	Risk identification is now part of normal business, automatic and ongoing.

(Continued on next page)

Table 5.3 (Continued)
Gold Coast Water Perspective on HACCP Application to Distribution System

Question	Answer	Basis/examples
Improved understanding of the distribution system?	Yes	The previous monitoring regime was based on population density and convenience. The HACCP system revealed weaknesses in smaller supply zones.
Improved understanding and maintenance of positive system pressure?	Yes	But not in the scope of Gold Coast HACCP.
Enhanced backflow and cross-connection prevention?	Yes	But these problems were adequately controlled before HACCP.
Has monitoring become more focused?	Yes	We now monitor according to where our risks are greater. The selection of sites, the frequency of tests and the selection of parameters are now based on risk, not just compliance. Customer complaints are better investigated and there is more accountability by distribution system operational staff.
Improved databases?	Yes	Central database has been altered to make investigation of trends easier.
Improved calibration schedules?	Yes	Calibration frequencies, record keeping and even techniques have been reviewed and improved. On-line instruments are checked against reference instruments during HACCP audits.
Better targeted reporting?	Yes	Reporting of all HACCP excursions is made to all senior management. Operational staff is given updates three times daily of current water quality complaints to help gauge impacts of their activities.
Improved record keeping?	Yes	Historical data with respect to water quality is superior in accessibility and scope now compared to pre-HACCP.
Made it easier for O&M staff to operate system?	Yes	HACCP excursion reports encourage best practice while simultaneously allowing operators to communicate design, maintenance, equipment, funding, and staffing issues to management.
Enhanced operator empowerment (ownership, understanding and involvement)?	Yes	Culture change has been marked. Water plant operators now would not personally consider operating above 0.2 NTU as acceptable. Reticulation staff now considers water quality impacts as part of the many activities they carry out. There is a high level of awareness that accountability is unavoidable and this drives ownership and involvement. Energy is required to drive this cultural environment. HACCP excursion reports encourage best practice while simultaneously allowing operators to communicate design, maintenance, equipment, funding, and staffing issues to management.

(Continued on next page)

Table 5.3 (Continued)
Gold Coast Water Perspective on HACCP Application to Distribution System

Question	Answer	Basis/examples
Did additional training in water safety get introduced?	Yes	HACCP procedures dictated the standard of performance required and gave instruction on methods to achieve this performance. This necessitated some "refresher" training e.g. jar testing skills, reservoir inspection techniques.
Did existing training become less ad-hoc and more structured?	N/A	Training was not a major emphasis of the HACCP Plan.
Improved senior management involvement?	Yes	In the Gold Coast model, management has no choice but to be intimately involved in the HACCP system. All HACCP excursions are sent to managers for their information. It would be rash of management to ignore these and in practice, they do not.
Improved relationships between management and operators?	Yes	HACCP excursion reports have in the past highlighted problems with staffing, training, asset condition, system design. Operators are happy that HACCP has brought attention to problems they wish addressed
Improved clarity of accountability?	Yes	Marked change in accountability through out the organization. HACCP system is designed to increase the probability that poor performance will be exposed. System has features that provide strong incentives for all levels of the organization to be accountable.
Has communication been improved within the utility and between the utility and external stakeholders?	Yes	HACCP has been such a success in this area that it has been extended to wastewater plants, effluent reuse, and trade waste management. Again, energy is required to keep communication flowing.
What are the estimated costs of implementation of the HACCP system?		In the Gold Coast example (2 water plants and 500,000 population) one full time tertiary qualified person is required.
What new processes and concepts did HACCP bring in?		Better monitoring and reporting. More rapid response to adverse conditions. Plant failures in turbidity, and manganese control have demonstrably improved. Water quality complaints are better controlled (numbers lower).
Was good use made of existing systems to minimize duplication and maximize integration?	Yes	The improvements have primarily come from doing a better job with what we have.

(Continued on next page)

Table 5.3 (Continued)
Gold Coast Water Perspective on HACCP Application to Distribution System

Question	Answer	Basis/examples
What is the estimated value of the benefits of these changes?		Gold Coast has a raw water supply that is low in risk and therefore, the consequences of failure are not necessarily serious. The reduction of water quality complaints by 50 percent in the last 4 years may or may not have a monetary value. The dramatic improvement in treated water turbidity since 2000 may or may not have a monetary value. HACCP has shown where capital and maintenance expenditure needs to be made and also, sometimes, where it does not need to be made e.g. replace filter backwash system in preference to repaint clarifier.
Has HACCP enhanced any environmental outcomes?	Yes	Gold Coast has been using HACCP in wastewater since August 2003. Data already shows decreases in the mass discharges of nitrogen and phosphorus from all three HACCP-certified plants.
How was the auditing (inspection, evaluation) process perceived?	Mixed reaction	Staff are now used to HACCP audits and understand their necessity and value. However, they are a little uncomfortable for staff because the audits are detailed. Trends are examined closely and cross checked with other information. Even on-line instruments are checked against references. The auditing is carried out by people with technical expertise. Utility staff is called on to account for any failures not properly reported.
Has continuous improvement been introduced?	Yes	Under the Gold Coast HACCP system, continuous improvement is unavoidable. In order to avoid HACCP excursions, operational staff focuses on being proactive. Management cannot be seen to be ignoring HACCP excursions in case the root cause is in their arena. The auditing system forces the communication both ways to continue. The "warning bell" nature of HACCP critical limits means improvement has a greater chance of being timely.
Has credibility and due diligence improved?	Yes	For almost every aspect of water quality management, there is a HACCP procedure designed to detail the risk, monitoring, corrective action and reporting. Gold Coast HACCP system monitoring results are available on request to customers and all HACCP internal audit reports and excursion notifications are "discoverable documents." The system is designed in the belief that at any time Gold Coast Water may have to produce proof of diligence. The regard for credibility and diligence is turned into a powerful driver by the structure of the HACCP system.

N/A – not applicable
* Water quality incident is defined as a major water quality problem leading to some kind of emergency response such as extensive dirty water events, taste and odor events or suspected contamination events.
HPCs – heterotrophic plate count bacteria

Table 5.4
South East Water Perspective on HACCP Application to Distribution System

Question	Answer	Basis/examples
Reduced taste/odor complaints?	Yes	Difficult to quantify, but overall, the number of water quality customer complaints per 1,000 customers has dropped by more than 60% since HACCP was introduced. However, all of this decrease cannot be attributed to HACCP. Some is due to changes in customer perceptions since water restrictions were introduced and changes in demand profiles with the introduction of water restrictions.
Reduced discolored water complaints?	Yes	See above.
Reduced low pressure complaints?	N/A	Not considered in HACCP plan.
Reduced *E. coli* or thermotolerant coliforms?	Yes	Number of samples with *E. coli* has reduced over time. This is partially due to HACCP and improved system understanding, and also due to increased disinfection levels.
Reduced total coliforms?	Yes	Numbers have decreased through better system understanding, some of which were related to HACCP.
Reduced HPCs?	N/A	Not considered in HACCP plan.
Reduced turbidity?	Yes	Turbidity-specific controls in place, although an improved understanding of the causes of turbidity is being used to minimize discolored water complaints.
Low pressures?	N/A	Not considered in HACCP plan.
Reduced water quality incidents?*	Yes	Water quality incidents have reduced from one every few months to none.
Reduced chlorinator failures?	N/A	No primary chlorinators in system.
Systematic identification and prioritization of risks and control?	Yes	Done as part of HACCP plan development and ongoing review.
Improved understanding of the distribution system?	Yes	As above.
Improved understanding and maintenance of positive system pressure?	Yes	Improved understanding of need for positive pressure. Primary distribution systems off gravity pressure systems.

(Continued on next page)

Table 5.4 (Continued)
South East Water Perspective on HACCP Application to Distribution System

Question	Answer	Basis/examples
Enhanced backflow and cross-connection prevention?	Yes	Improvements made to backflow prevention strategy.
Has monitoring become more focused?	Yes	Operational monitoring was more focused at understanding causes for events. However, this is not directly driven by HACCP.
Improved databases?	Yes	This has occurred from continuous improvement and not HACCP.
Improved calibration schedules?	Yes	This has been driven by the HACCP review process.
Better targeted reporting?	Yes	More for internal business reporting than HACCP.
Improved record keeping?	Yes	Directly as a result of HACCP, improved logging of all system check records.
Made it easier for O&M staff to operate system?	Yes	Training of O&M staff has improved their actions, further improving our water quality.
Enhanced operator empowerment (ownership, understanding and involvement)?	Yes	Ability for operators to have a say and have recommended changes implemented.
Did additional training in water safety get introduced?	Yes	Yes, through toolbox meetings with external contractor and internally through in-house training of all operations, planning and strategic staff.
Did existing training become less ad-hoc and more structured?	Yes	Through the development of dedicated programs targeted around HACCP.
Improved senior management involvement?	Yes	Through greater awareness of HACCP and internal training.
Improved relationships between management and operators?	Maybe	Relationships already good prior to introduction of HACCP.
Improved clarity of accountability?	Yes	Particularly as identified through record keeping procedures.

(Continued on next page)

Table 5.4 (Continued)
South East Water Perspective on HACCP Application to Distribution System

Question	Answer	Basis/examples
Has communication been improved within the utility and between the utility and external stakeholders?	Yes	This is only partially through HACCP, more through alliance partnering.
What are the estimated costs of implementation of the HACCP system?		Of the order \$80,000-\$120,000 (Australian Dollars) (\$65,000-\$95,000 US Dollars) initially, this includes costs for in-house staff time in actual compilation of the plan.
What new processes and concepts did HACCP bring in?		HACCP and the risk management framework, introducing proactive management to address or prevent risks.
Was good use made of existing systems to minimize duplication and maximize integration?	Yes	The existing quality documentation system was used as the backbone for HACCP.
What is the estimated value of the benefits of these changes?		Unable to be quantified. There are fewer complaints, less incidents, and no major health impacts from the water supply system with estimated internal savings on the order of \$100,000-200,000 Australian Dollars (\$80,000-160,000 US Dollars) per year. But how is a value put on increased public and regulator confidence? It is not possible to conclusively distinguish which savings arose from HACCP directly or the improved system understanding, training, and other factors.
Has HACCP enhanced any environmental outcomes?	No	
How was the auditing (inspection, evaluation) process perceived?		As an education experience.
Has continuous improvement been introduced?	Yes	This has been an integral component of the organization prior to HACCP.
Has credibility and due diligence improved?	Yes	Through improvements in record keeping.

* Water quality incident is defined as a major water quality problem leading to some kind of emergency response such as extensive dirty water events, taste and odor events or suspected contamination events.

Table 5.5
Yarra Valley Water Perspective on HACCP Application to Distribution System

Question	Answer	Basis/examples
Reduced taste/odor complaints?	No	Complaints data.
Reduced discolored water complaints?	No	Complaints data. Factors such as drought recovery affected complaints numbers and are not a good indicator to assess effectiveness of HACCP.
Reduced low pressure complaints?	N/A	Not part of HACCP.
Reduced *E. coli* or thermotolerant coliforms?	Maybe	Compliance improved every year since 1996. Cannot attribute totally to implementations of HACCP.
Reduced total coliforms?	Maybe	Compliance improved every year since 1996. Cannot attribute totally to implementations of HACCP.
Reduced HPCs?	N/A	Not part of HACCP.
Reduced turbidity?	No	Water quality data.
Reduced low pressures?	N/A	Not part of HACCP.
Reduced water quality incidents?*	Yes	Security breaches at storages minimized due to improved processes.
Reduced chlorinator failures?	Yes	Significant improvement due to joint Melbourne Water and Yarra Valley Water HACCP Management via regular HACCP coordinator meetings and reporting processes.
Systematic identification and prioritization of risks and control?	Yes	Via ongoing HACCP management and improvements identified during internal and external audits.
Improved understanding of the distribution system?	Yes	Development of a risk management plan for hydrant use.
Improved understanding and maintenance of positive system pressure?	Yes	Appointment of a Backflow Prevention Officer. Implementation of significant improvements to the backflow prevention program and allocation of budget via the annual business planning process.
Enhanced backflow and cross-connection prevention?	Yes	As above.

(Continued on next page)

Table 5.5 (Continued)

Yarra Valley Water Perspective on HACCP Application to Distribution System

Question	Answer	Basis/examples
Has monitoring become more focused?	Yes	External review of annual water quality monitoring program identified areas of improvements based on risk assessment in the 2004 Australian Drinking Water Guideline Framework and changes introduced accordingly.
Improved databases?	Yes	Significant improvements to the database to undertake monthly checks for HACCP management via automated reports.
Improved calibration schedules?	N/A	
Better targeted reporting?	Yes	Monthly water quality report to check compliance against HAACP criteria and undertake investigations or remedial works. Monthly water quality results for each water quality zone posted in the internet. Improvements identified via HACCP reviews and audits are captured, progress monitored and reported.
Improved record keeping?	Yes	As above.
Made it easier for O&M staff to operate system?	Maybe	Improved awareness of water quality risks. Improvements to the water quality Emergency Response Plan.
Enhanced operator empowerment (ownership, understanding and involvement)?	No	Well-established processes were in place prior to HACCP and ongoing continuous improvement of processes.
Did additional training in water safety get introduced?	Yes	Annual training of key staff including contactors on HACCP and Water Quality.
Did existing training become less ad-hoc and more structured?	Yes	As above.
Improved senior management involvement?	Yes	Improvements identified via HACCP reviews and audits are captured, progress monitored and reported via an automated monitoring process in the intranet. The progress reported to Senior Management via the intranet and also regular Quality Council Committee meetings.
Improved relationships between management and operators?	N/A	
Improved clarity of accountability?	N/A	
Has communication been improved within the utility and between the utility and external	Yes	Significant improvement of communication and critical control point management with the wholesaler.

(Continued on next page)

Table 5.5 (Continued)

Yarra Valley Water Perspective on HACCP Application to Distribution System

Question	Answer	Basis/examples
stakeholders?		
What are the estimated costs of implementation of the HACCP system?		$10,000 (Australian Dollars) ($8,000 US Dollars) for annual HACCP reviews, internal and external audits etc.
What new processes and concepts did HACCP bring in?		Improved backflow prevention, disinfection failure management.
Was good use made of existing systems to minimize duplication and maximize integration?	Yes	Existing business management system comprising ISO 9000 and ISO 14000 and emergency management plans and documented quality procedures.
What is the estimated value of the benefits of these changes?		Reduction in responses for disinfection failures and potential backflow.
Has HACCP enhanced any environmental outcomes?	Yes	HACCP now being used to manage recycled water quality management.
How was the auditing (inspection, evaluation) process perceived?	Positive	A system audit process was already in place before HACCP. Many suggestions for improvements from HACCP audits were implemented.
Has continuous improvement been introduced?	Yes	This was already in place and HACCP improvements were incorporated to the existing process. The progress of improvements are managed and monitored via an automated tracking process via the intranet.
Has credibility and due diligence improved?	Yes	Improved awareness on water quality risks and management processes.

* Water quality incident is defined as a major water quality problem leading to some kind of emergency response such as extensive dirty water events, taste and odor events or suspected contamination events.

Reductions in Customer Complaints

The most notable improvement reported by utilities following implementation of HACCP was an apparent drop in the frequency of customer complaints. For example, water quality complaints dropped steadily following HACCP implementation for Gold Coast Water and South East Water by a total of 66 percent and 60 percent, respectively, over a few years. In the case of Gold Coast Water, this reduction in complaints occurred at a time of increased veracity in capturing and recording complaints, such that the true magnitude of reduction may have been greater. For South East Water, interpreting the cause for this reduction was complicated because of changes in demand and customer expectations associated with water restrictions and an ongoing drought.

Improvements in Water Quality

A number of specific water quality improvements were attributed, at least partly, to HACCP implementation. Case study utilities reported improvements against a range of criteria as follows:

- Microbial indicators: utilities reported improved compliance with microbial results, although this started from a relatively high baseline, so changes were not large. Yarra Valley Water reported ongoing, year-by-year, compliance improvements attributed to many factors of which HACCP implementation was one. Gold Coast Water reported improved compliance from 2 percent to 1.5 percent for thermotolerant coliform occurrences.
- Incidents: water quality incidents were reported to have been reduced for South East Water who reported a drop from one incident every few months to less than one per year. A water quality incident is defined as a major water quality problem leading to some kind of emergency response, such as extensive dirty water events, or taste and odor events.

Improvements in Process Performance

Although outside of the scope of the distribution component of the project, it is worth noting a number of process performance improvements attributed, at least partly, to HACCP implementation:

- Filtration plant: Filtration plant performance has been improved following the implementation of HACCP. For example, since the Gold Coast Water HACCP system was put into place in 2001, the water treatment and distribution system performance parameters have seen steady improvement (Smith 2004a). The average weekly turbidity of the finished water has improved from 0.28 NTU in July 1999 to 0.18 NTU in June 2003, as shown in Figure 5.2.
- Disinfection plant: Chlorinator performance has been improved sine the implementation of HACCP. For example, Yarra Valley Water saw a steady decline in the number of chlorinator critical control point exceedances from around twenty per year to one per year during a four-year period following HACCP implementation as shown in Figure 5.3 (Chapman, Jayaratne, and Pamminger 2003).

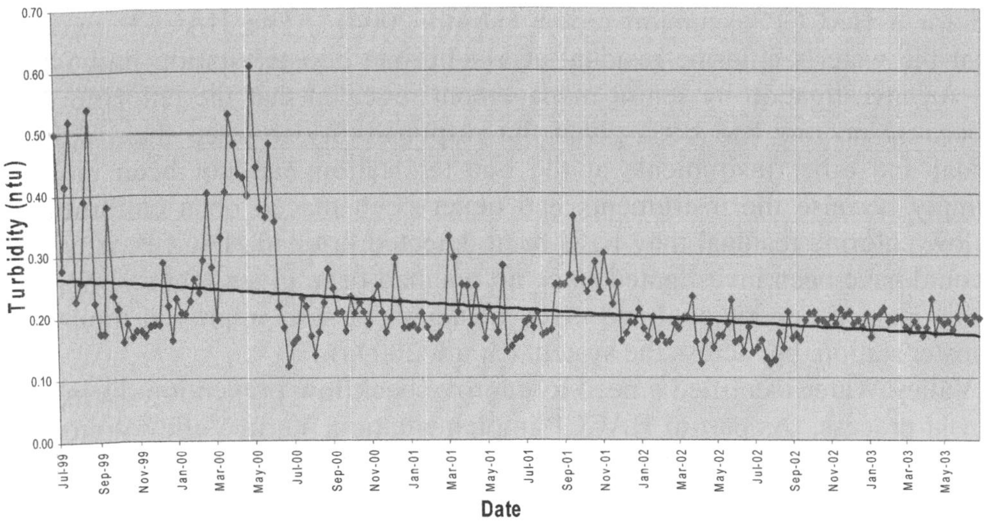

Source: Smith 2005. Reprinted with permission.
ntu = nephalometric turbidity units

Figure 5.2 Gold Coast Water average weekly turbidity of finished water from Molendinar Water Treatment Plant

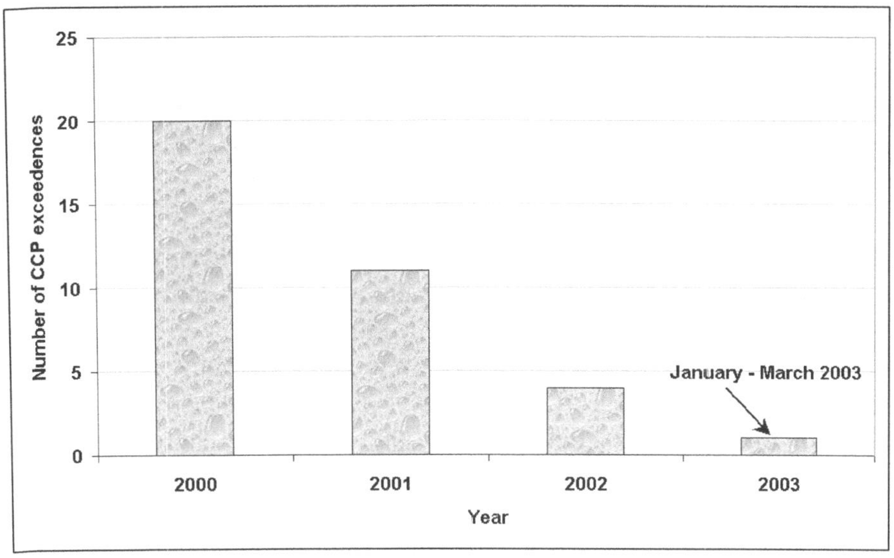

Source: Chapman, Jayaratne, and Pamminger 2003. Reprinted with permission from AWWA.
CCP = critical control point

Figure 5.3 Number of chlorinator CCP exceedances per year

Improvements in Work Processes

Gold Coast Water indicated that certain operational practices were improved based on data collected for a HACCP excursion report (Smith 2004). The HACCP excursion report established that the water's chlorine residual at a suburban booster station had fallen below a critical limit. An investigation by senior management revealed that the chlorine gas cylinders were empty because no one had been given the responsibility to keep them full. Also, the chlorine residual and other instruments at the booster station had not been calibrated since installation simply because the instruments had never been placed on a calibration schedule. Although the low chlorine residual may have been detected prior to HACCP, it is unlikely that the incident would have been investigated since no one had been given that responsibility. As a follow-up to this incident, operational and asset management staff improved their practices not only at this booster station, but across the system (Smith 2004a).

Yarra Valley Water identified a need to improve backflow prevention during its HACCP plan development process. As part of HACCP implementation, Yarra Valley Water appointed a dedicated Backflow Prevention Officer. With the officer coordinating and managing this important issue, the utility is now aiming to increase the number of connections that have the appropriate backflow prevention devices fitted as part of a retrofit program with a target of 100 percent compliance with the relevant backflow prevention standard within five years.

Improved Understanding of Risks and Risk Management

Smith (2004a) documents a case study at Gold Coast Water in Australia that illustrates how a HACCP excursion report helped the utility to identify a previously unidentified risk in an existing manganese control procedure and its reporting instructions. "The loophole in the HACCP manganese control procedure meant that staff did not technically have to report this particular failure to anyone and this became an attractive option" (Smith 2004a). The HACCP excursion reporting system was developed as part of the HACCP Plan, specifically addressing HACCP step 12, Documentation and Recordkeeping. The HACCP excursion report showed that the daily limit of six customer calls regarding dirty water (Smith 2003b) had been exceeded on one day. A follow-up investigation conducted in response to the high number of dirty water calls revealed a problem where plant staff failed to adequately respond to changes in raw water quality. The existing procedure was subsequently tightened.

In another example, at Gold Coast Water's Mudgeeraba plant, Smith (2003b) reports that there was initially under-reporting of a turbidity problem due to inferior monitoring (one combined meter compared to six meters at another plant). There was also more cultural resistance to compliance at this plant and it took a change in instrumentation and some close internal auditing to reveal issues at that plant in 2002. Two problems were revealed. One was that operators were inexperienced in dosage optimization during changing conditions. Secondly, the backwash system components were in poor condition which compromised filter performance under certain conditions. The lack of cultural acceptance was dealt with by closer internal auditing in the early stages and eventually staff realized that there were benefits all around if everyone adhered to the system (Smith 2003b).

Improvements in Documentation and Record Keeping

Brisbane Water and Gold Coast Water noted that HACCP implementation led to improved documentation and fine-tuning of existing procedures. For example, Gold Coast Water (Smith 2004a) had no record of system failures prior to HACCP implementation (Smith 2003b). A HACCP excursion reporting system was implemented as part of the HACCP Plan. In December 2002, the Health Dcpartment advised Gold Coast Water of a small outbreak of cryptosporidiosis in one part of the city. The Health Department needed to know if this outbreak was caused by the public water supply. Within 20 minutes of receiving the call, the utility was able to collate comprehensive data records that indicated the outbreak was not likely caused by the water supply. These records included weekly counts for *E. coli* at the source of supply, 24-hour trend data for various water quality parameters (e.g. raw water turbidity, dosed water pH, filtered water turbidity, plant chlorine residuals), supply reservoir inspection reports, weekly water quality data for a sampling location adjacent to the area of concern, and consumer call records. The utility had a high degree of confidence in the information provided to the Health Department because the sampling and laboratory work was carried out by a certified lab and the trended data was provided by instruments with up-to-date calibration records and verified several times daily against a reference instrument (Smith 2004a).

Demonstration of Due Diligence

Davison and Deere (2004) illustrate how HACCP can help a utility to demonstrate "due diligence" or the "prevention of foreseeable harm." This demonstration of due diligence is accomplished using five elements of the HACCP system (Davison and Deere 2004): (1) assessing risks from the sources of supply through to the customer's tap; (2) implementing a risk management system; (3) employing a "culture of compliance;" (4) seeking out and incorporating new knowledge into system processes and procedures; and (5) planning for emergencies.

South East Water, Yarra Valley and Gold Coast Water found that HACCP implementation has helped to improve system credibility and demonstration of due diligence. South East Water attributed these improvements to improved record-keeping practices. For Yarra Valley, employees have an improved awareness of water quality risks and management processes. Gold Coast Water found that the structure of the HACCP system has helped with system credibility and demonstration of due diligence. For almost every aspect of water quality management, Gold Coast Water has a HACCP procedure designed to detail the risk, monitoring, corrective action, and reporting. Each procedure is a formal instruction from the manager to the operational staff (Smith 2004a). These procedures, typically one to three pages in length, outline the process steps, the risks, the monitoring strategies, control measures, corrective actions, and assignment of responsibilities.

SUMMARY

Important conclusions have emerged from analyzing the five case studies presented in this chapter:

- Most utilities that have gained HACCP certification have done so after some core management systems (e.g. ISO 9001, ISO 14001) had been developed and implemented. These management systems helped the utility to gain management control of people and processes. For example, the additional documentation and record-keeping requirements of HACCP are often quite small once an organization has experience with a management system.
- In practice, the case study utilities did not operate multiple, separate systems for quality management, occupational safety, water quality and safety, and environmental considerations. These separate systems were captured, as required, within an integrated management system (IMS). The principal benefit of an IMS, as identified by water utilities, was the avoidance of duplication, leading to reduced staff time and costs, and improved process integration.
- Auditing, although sometimes uncomfortable for operating staff, is a necessary and useful element of HACCP. It forces periodic reviews and keeps important issues before staff and management.
- When a utility implements HACCP for water, it is likely to be doing so as one of a number of water quality improvement projects, making it very difficult, if not impossible, to precisely determine the extent to which HACCP per se contributed to any effect.
- Improvements did become evident following the implementation of HACCP, but in most cases, those changes did not appear conclusive until a consistent pattern of improvement had emerged after three or more years.

These conclusions have major implications for the pilot-studies undertaken by four utilities as part of this Project, as well as implications beyond this study and would apply to any project that attempts to measure the effect of the implementation of HACCP. Given these limitations, conclusive evidence of major water quality or other improvements resulting directly from the implementation of HACCP would not be expected to be observable in the timeframe, or within the context of this study as reported in Chapter 4.

Importantly, a utility that voluntarily implements a certified HACCP plan would most likely not do so unless key staff had some reasonable confidence in the soundness of their water quality management practices and a prevailing belief that water quality management is important. Such utilities are already likely to be managing water quality at a higher level than those that would reject the implementation of HACCP providing less opportunity for improvement. The relatively limited improvements reported by City West Water and Brisbane Water were, for example, attributed by those utilities to the fact that by the time they implemented HACCP, most of the requirements for HACCP implementation were already in place, including mature certified quality management systems.

Despite these study limitations, the five case study utilities all reported some benefits in aspects of water quality management that all believed were at least partly related to HACCP. Furthermore, none of these utilities have withdrawn from their voluntary certification programs.

Improvements considered by utilities to be associated with the implementation of HACCP included the following:

Water quality

- Reduced customer complaints
- Reduced water quality incidents
- Reduced microbial indicator failures

Operations

- Improved treatment system performance
- Improved work processes.

Business performance

- Improved capability to demonstrate due diligence
- Reduced costs
- Improved documentation and record keeping
- Improved training and operational good practice culture.
-

Since the HACCP implementation costs for these five utilities were relatively modest, and in some cases were met entirely with existing resources, it appears that implementation of HACCP may represent a positive overall experience for mid- to large-sized water utilities. However, it is important to note that cost information presented in this chapter represents utilities that had implemented other quality management systems prior to implementing HACCP and therefore, the cost information may not be applicable to a utility implementing only HACCP.

The results of the analysis presented in this Chapter are not necessarily transferable to very small water utilities that may lack the critical mass of funding, expertise and staff numbers to undertake a HACCP program. No very small water utilities with HACCP certification could be identified to be used as case studies for this Chapter. Referring back to Chapter 4, some small systems were considered through the pilot utilities. However, although the Power and Water Corporation Katherine pilot-study demonstrated the application of HACCP to a small system, it should be noted that the Katherine water supply system is managed by a mid-size water authority. The South Berwick Water District in Maine, a small system serving approximately 4,000 customers, was unable to implement its HACCP plan due to manpower limitations and the need to focus on higher priority projects.

CHAPTER 6
RECOMMENDED NEXT STEPS AND FUTURE RESEARCH NEEDS

The HACCP system has been widely adopted by the food and beverage industries, and has been incorporated in many drinking water regulatory requirements and guidelines around the globe. For example, the third edition of the WHO drinking water guidelines (WHO 2004) outlines a framework for drinking water safety based on the multiple barrier approach and several risk management and quality management approaches including HACCP.

Many European and Australian water utilities have implemented HACCP systems to full certification or have implemented approaches that are more or less equivalent, albeit differently labeled. The US drinking water industry has been slower than its counterparts overseas in applying HACCP or similar approaches to watershed management programs, water treatment facilities and distribution systems. The US drinking water community can use the experiences from European and Australian utilities as a practical basis for applying the HACCP principles to improve protection of the water distribution system. Although several HACCP applications in Australia and Ontario, Canada were motivated by waterborne disease outbreaks or suspected contamination incidents, a superior approach would be to develop and implement quality assurance programs such as HACCP prior to experiencing such an outbreak or contamination event.

A recommended first step is to improve utility programs and practices to meet the AWWA Standard G200-Distribution Systems Operation and Maintenance. The specifications for this standard include many water utility programs and practices that are needed to successfully implement HACCP. Examples of supporting programs include an emergency response program, a water main flushing program, and a tank inspection program. Examples of specific SOPs for the water distribution system may include water quality sampling procedures, calibration procedures for on-line monitoring devices, and security procedures for restricting entry of unauthorized personnel. The distribution system HACCP plan will simply include and reference these programs and practices to avoid duplicative work. The programs/practices may already exist in some form, but may need updating or augmentation.

Other recommended steps for water utilities include:

- Consider developing and implementing the ISO management systems (ISO 9001 and ISO 14001) or an integrated management system that includes HACCP as well as the ISO systems.
- Evaluate HACCP applications to other water supply system components.

Additional recommended action items for the drinking water community include:

- Consider the HACCP framework in US regulations or regulatory guidelines.
- Evaluate HACCP applications in small water systems.

These recommended steps and actions items are discussed in the following sections.

Consider Developing and Implementing the ISO Management Systems

The case studies discussed in Chapter 5 showed that most utilities that have gained HACCP certification have done so after some core management systems (e.g. ISO 9001, ISO 14001) had been developed and implemented. These management systems helped the utility to gain management control of people and processes. Also, the case study utilities did not operate multiple, separate systems for quality management, occupational safety, water quality and safety, and environmental considerations but captured these systems within an integrated management system.

Evaluate HACCP Applications to Other Water Supply System Components

Some of the participating utilities have implemented comprehensive HACCP systems, from source of supply to connection. In addition, one of the project Principal Investigators, from Melbourne Water, has developed a watershed HACCP plan for watershed and reservoir management. Water utilities are encouraged to consider applying HACCP principles to watershed management programs and treatment facilities in addition to the distribution system. This investigation may be a joint effort amongst the utility, USEPA regulators, state regulators and other stakeholders.

Consider HACCP Framework in US Regulations or Regulatory Guidelines

The USEPA is currently revising the Total Coliform Rule and is considering new possible distribution system requirements as part of these revisions. As part of this process, the USEPA is currently developing a series of issue papers on topics relevant to the Rule. An issue paper on the HACCP system has been developed as part of this effort. The next step for the USEPA is to present and discuss these issue papers at an expert workshop to provide further refinement and to support any rulemaking activity. To support USEPA's evaluation of HACCP, it would be helpful to have additional US case studies on HACCP applied to water distribution systems.

Evaluate HACCP Applications in Small Water Systems

Further research is needed to document the costs and benefits of implementing HACCP by small water systems and to answer other outstanding questions, as listed below. This research is needed to develop better guidance information for small systems, and to help regulators evaluate HACCP as a possible tool or framework as part of upcoming regulatory developments. No small water utilities with HACCP certification could be identified to be used as case studies for this Project.

More information is needed to assess the adequacy of supporting programs and practices in small systems. Small systems may rely on verbal information exchanges rather than written SOPs. While this may be acceptable for many situations, there can be significant risks if too much reliance is placed on procedures that are not systematic, formalized and documented. When key staff is on vacation or becomes sick, or if there is no opportunity for checking and auditing, it is difficult for health authorities to be confident that water quality risks are being reliably managed. Internal auditing, employee training programs, document control and data management systems require additional resources that may not be available in small systems at present. This means that small water utilities are potentially under resourced or under supported by state and federal agencies.

Outstanding questions on HACCP implementation in small systems include:

- How should small systems develop a HACCP Plan (outside consultant vs. in-house HACCP team; use of guidelines or checklists; workshop setting; committee meetings)?
- How much training on HACCP is needed for system employees?
- What is the man-hour requirement for developing and implementing a HACCP Plan?
- What are the documented costs and benefits of implementing HACCP for small systems?
- What are the barriers to implementing HACCP in a small system, and what are the strategies to overcome these barriers (e.g. multiple demands on resources)?
- Does the small system have adequate historical water quality and system data to complete the HACCP hazard analysis (Step 6)?
- Does the small system have the appropriate utility culture in place to be successful in implementing HACCP?

REFERENCES

ActewAGL. 2005. HACCP Plan Working Draft Version 5.0. Internal Document, Canberra, Australia: ACTEWC.

AWWA. 2005. Three possible accreditation standards piloted. *E-MainStream*, 2(3):1.

AWWA. 2004. *ANSI/AWWA G200-04 Standard - Distribution Systems Operation and Management*. Denver, Colo.: AWWA.

AWWA. 2001. Emergency Response Planning for Water Utilities. Manual of Water Supply Practices #19, Fourth Edition. Denver, Colo.: AWWA.

AWWA (American Water Works Association). 2001a. *Accreditation Manual of Policies and Procedures – Draft.* AWWA. Denver, CO.

AwwaRF (American Water Works Association Research Foundation). 2005. Project Abstract – Developing an Environmental Management System for Water Utilities. www.awwarf.com/research/TopicsAndProjects/abstract.aspx?pn=2930.

Barry, S.J., E.R. Atwill, K.W. Tate, T.S. Koopman, J. Cullor, and T. Huff. 1998. Developing and Implementing a HACCP-based Program to Control *Cryptosporidium* and Other Waterborne Pathogens in the Almeda Creek Watershed: Case Study. In *Proc. of the AWWA Annual Conference*. Denver, Colo.: AWWA.

Barry, S.J., T.S. Koopmann, T. Huff, E.R. Atwill, K.W. Tate, and J. Cullor. 2004. A HACCP-based Program to Protect San Francisco's Drinking Water. In *Proc. of the NSF Conference on Risk Management Strategies for Drinking Water Utilities*. www.nsf.org/cphe/cphe_rms_presentations.html

Bosshart, U. 2003. HACCP – Hazard Analysis and Critical Control Points at the Zurich Water Supply. In *HACCP in Drinking Water Supplies in Switzerland*. Zurich, Switzerland: Swiss Gas and Water Industry Association.

Breach, B. 2004. The Bonn Workshops – A Charter for Safe Drinking Water. In *Proc. of the NSF Conference on Risk Management Strategies for Drinking Water Utilities*. www.nsf.org/cphe/cphe_rms_presentations.html

Bryan, J.J. 1993. Hazard Analysis and Critical Control Points and Their Application to the Drinking Water Treatment Process. In *Proc. of the 1993 AWWA Water Quality Technology Conference*. Denver, Colo.: AWWA.

Chapman, T., A. Jayaratne, and F. Pamminger. 2003. Water Quality Benefits Gained From Having a Hazard Analysis and Critical Control Point (HACCP) System in Place Since 1999. In *Proc. of the AWWA Annual Conference*. Denver, Colo.: AWWA.

Clark, Robert M., D.A. Reasoner, K.R. Fox, and C.J. Hurst. 1997. Drinking Water Treatment and Distribution: Its Role in Preventing Waterborne Outbreaks. Presented at Workshop on Design of Waterborne Disease Occurrence Studies. March 12-13, 1997, Atlanta, GA.

Codex. 1997. *Guidelines for the Application of the HACCP System*. Rome, Italy: Codex Alimentarius Commission and the FAO/WHO Food Standards Program, Food and Agriculture Organization of the United Nations and World Health Organization.

Codex Alimentarius Commission. 1993. *Guidelines for the Application of the Hazard Analysis Critical Control Point (HACCP) System*, CAC/GL 18-1993.

Cotruvo, J. 2004. Comprehensive Water Quality Management Systems: Potential Scope of a Complete System. In *Proc. of the NSF Conference on Risk Management Strategies for Drinking Water Utilities*. www.nsf.org/cphe/cphe_rms_presentations.html

Cunliffe, D.A. 2004. Australian Framework for Management of Drinking Water Quality. In *Proc. of* the *NSF Conference on Risk Management Strategies for Drinking Water Utilities*. www.nsf.org/cphe/cphe_rms_presentations.html

Cunliffe, D.A. 2001. Australia's Framework for Management of Drinking Water Quality. In *Proc. of the AWWA Annual Conference*. Denver, Colo.: AWWA.

Davison, A., and J. Bartram. 2004. HACCP and WHO Water Safety Plans. In *Proc. of the NSF Conference on Risk Management Strategies for Drinking Water Utilities*. www.nsf.org/cphe/cphe_rms_presentations.html

Davison, A., S. Davis, and D. Deere. 1999. Quality Assurance and Due Diligence for Water: Can HACCP Deliver? In *Proc. of the Australian Water and Wastewater Association Conference*.

Davison, A. and D. A. Deere. 2004. Liabilities, Due Diligence and Utilities: HACCP as a Legal Defense. In *Proc. of the NSF Conference on Risk Management Strategies for Drinking Water Utilities*. www.nsf.org/cphe/cphe_rms_presentations.html

Davison, A. and D. A. Deere. 1999. Safety on Tap! *Microbiology Australia*, 20(2):28-31.

DeBier, L. and J-C. Joret. 2004. HACCP as a Tool for Assuring Safe Drinking Water: The Veolia Water Experience. In *Proc. of the NSF Conference on Risk Management Strategies for Drinking Water Utilities*. www.nsf.org/cphe/cphe_rms_presentations.html

Deere, D. 2005. Question on Sydney Water. Email to K. Martel. 1.3.05.

Deere, D. 2004. Implementation of a HACCP Approach to Watershed Management. In *Proc. of the NSF Conference on Risk Management Strategies for Drinking Water Utilities*. www.nsf.org/cphe/cphe_rms_presentations.html

Deere, D.A. 1999. Applicability of HACCP in the Water Industry. In *Proc. of the Conference Harmonized Risk Assessment for Water-Related Microbiological Hazards*. Geneva, Switzerland: World Health Organization.

Deere, D. and A. Davison. 1999. Assuring Water Quality. *WATER21*, 1(1): 8-9.

Deere, D. and A. Davison. 1998. Safe Drinking Water: Are Food Guidelines the Answer? *WATER*:Nov/Dec.

Deere, D., M. Stevens, A. Davison, G. Helm, and A. Dufour. 2001. Chapter 12 - Management Strategies. In: *Water Quality: Guidelines, Standards and Health. Assessment of Risk and Risk Management for Water-Related Infectious Disease*. London, England: IWA Publishing.

Deighton-Smith, R. and B. Labza. 2005. Determining benefits of a regulatory framework for drinking water quality. *Journal of the Australian Water Association*, 32(2):94-103.

Fok, N. and K. Emde. 2004. HACCP in Water Treatment – What It Is and What It Isn't. In *Proc. of* the *NSF Conference on Risk Management Strategies for Drinking Water Utilities*. www.nsf.org/cphe/cphe_rms_presentations.html

Friedman, M.J., G. Kirmeyer, G. Pierson, S. Harrison, K. Martel, A. Sandvig, and A. Hanson. 2005. *Development of Distribution System Water Quality Optimization Plans*. Denver, Colo.: AwwaRF.

Friedman, M.J., L. Radder, S. Harrison, D. Howie, M. Britton, G. Boyd, H. Wang, R. Gullick, M. LeChevallier, J. Funk and D. Wood. 2004. *Verification and Control of Low Pressure Transients in Distribution Systems*. Denver, Colo.: AwwaRF and AWWA.

Geldreich, E.E. 1996. *Microbial Quality of Water Supply in Distribution Systems*. Boca Raton, FL: Lewis Publishers.

Gissurarson, L.R. 2004. HACCP and ISO Systems at Reykjavik Energy. In *Proc. of* the *NSF Conference on Risk Management Strategies for Drinking Water Utilities*. www.nsf.org/cphe/cphe_rms_presentations.html

Gray, R. and M. Morain. 2000. HACCP Application to Brisbane Water. *Water*, 27: 41-43.

Havelaar, A.H. 1994. Application of HACCP to Drinking Water Supply. *Food Control*, 5(3):145-152.

Hebenstreit, C. 2004. EMS Initiatives in the State of Missouri. In *Proc. of the NSF Conference on Risk Management Strategies for Drinking Water Utilities*. www.nsf.org/cphe/cphe_rms_presentations.html

Hrudey, S.E. and R. Walker. 2005. Walkerton 5 Years Later – Tragedy Could Have Been Prevented. *Opflow*, 31(6):1, 4-7.

International Studies and Training Institute, Inc. 1995. *Final Report: Sanitary Survey of the Drinking Water Distribution System of the District of Columbia*. Whiteford, MD: ISTI.

Kirmeyer, G.J., M. Friedman, J. Clement, A. Sandvig, P.F. Noran, K. Martel, D. Smith, M. LeChevallier, C. Volk, E. Antoun, D. Hiltebrand, J. Dykesen, and R. Cushing. 2000. *Guidance Manual to Maintain Distribution System Water Quality*. Denver, Colo.: AwwaRF and AWWA.

Kirmeyer, G.J., M. Friedman, K. Martel, D. Howie, M. LeChevallier, M. Abbaszadegan, M. Karim, J. Funk and J. Harbour. 2001. *Pathogen Intrusion Into the Distribution System*. Denver, Colo.: AwwaRF and AWWA.

Kirmeyer, G.J., L. Kirby, B.M. Murphy, P.F. Noran, K. Martel, T.W. Lund, J.L. Anderson, and R. Medhurst. 1999. *Maintaining and Operating Finished Water Storage Facilities to Prevent Water Quality Deterioration*. Denver, Colo.: AwwaRF and AWWA.

Kramer, M.H., B.L. Herwaldt, G.F. Craun, R.L. Calderon, and D.D. Juranek. 1996. Waterborne Disease: 1993 and 1994. *JAWWA*, 88(3):66-80.

Kuslikis, B. and B. White. 2004. Implementing a Sustainable, Integrated Risk Management System for Water Safety and Quality. In *Proc. of the NSF Conference on Risk Management Strategies for Drinking Water Utilities*. www.nsf.org/cphe/pdf/3Hebenstreit.pdf

Lee, J.J., P. Schwartz, P. Sylvester, L. Crane, J. Haw, H. Chang and H.J. Kwon. 2003. *Impacts of Cross Connections in North American Water Supplies*. Denver, Colo.: AwwaRF and AWWA.

Levy, D.A., M.S. Bens, G.F. Craun, R.L. Calderon, B.L. Herwaldt. 1998. Surveillance for Waterborne Disease Outbreaks--United States, 1995-1996. www.cdc.gov/epo/mmwr/preview.

Lippy, E.C. and, S.C. Waltrip. 1984. Waterborne Disease Outbreaks - 1946-1980: A Thirty-Five Year Perspective. *JAWWA,* 76:60-67.

Mays, L.W. 2004. Chapter 1 Water Supply Security: An Introduction. In *Water Supply Systems Security*. New York, NY: McGraw-Hill Books, Inc.

Melbourne Water Corporation. 2001. HACCP Plan (Version 3). Internal Document. Melbourne, Australia: Melbourne Water Corporation.

Metge, S. 2003. The French Experience with the Application of HACCP Principles in Drinking Water. In *Water Safety-Risk Management Strategies for Drinking Water - Conference Materials*. Berlin, Germany: Federal Environmental Agency (Umweltbundesamt).

Ministry of Health. 2001. *How to Prepare and Develop Public Health Risk Management Plans for Drinking-Water Supplies*. Wellington, New Zealand: Ministry of Health.

Mucklow, R. 1997. Where Did HACCP Come From? In *Heads Up for HACCP*. National Meat Association. www.nmaonline.org/files/headsup12-1.htm , November 3, 2004.

Mullenger, J. 2002. HACCP Principles 6 and 7. In *The How, Where and Why of Applying HACCP to Water – AWWA WQTC Workshop Manual*. Melbourne, Australia: Melbourne Water Corporation.

Nadeau, M. 2004. Letter to K. Martel (July). South Berwick, Maine: South Berwick Water District.

NASA (National Aeronautics and Space Administration). 1991. A Dividend in Food Safety. *Spinoff 1991*, NASA Technical Report ID 20020086314.

NHMRC/NRMMC (National Health and Medical Research Council and Natural Resource Management Ministerial Council). 2004. *Australian Drinking Water Guidelines*. Canberra, Australian Capital Territory, Australia: National Health and Medical Research Council and Natural Resource Management Ministerial Council.

Nokes, C. 2001. Public Health Risk Management Plans: An Introduction to New Zealand's Use of a Risk Management-Based Approach to Improving the Safety of Drinking-Water Supplies. In *Proc. Of the NZIEH Conference*.

Nyman, L. 2004. Benefits of ISO Programs at Elizabethtown Water Company. In *Proc. of the NSF Conference on Risk Management Strategies for Drinking Water Utilities*. www.nsf.org/cphe/cphe_rms_presentations.html

O'Connor, D.R. 2002. *Report of the Walkerton Inquiry. Part 2. A Strategy for Safe Drinking Water*. Toronto, Canada: The Walkerton Inquiry.

Shamsuddin, A.J., S.G. Mahmud, M.F. Ahmed, D. Deere, A. Davison, and G. Howard. 2005. Developing a water-safety framework in Bangladesh. *Waterlines,* 23(4): 1118.

Smith, D. 2005. Updated turbidity graph. Email to D. Deere. 8.19.05.

Smith, D. 2004a. HACCP in Water: The War on Error. In *Proc. of the NSF Conference on Risk Management Strategies for Drinking Water Utilities*. www.nsf.org/cphe/cphe_rms_presentations.html

Smith, D. 2004b. Updated turbidity and complaints graphs. Email to K. Martel. 6.8.04.

Smith, D. 2003a. HACCP Procedures. Email to K. Martel. 2.17.03.

Smith, D. 2003b. HACCP Case Study. Email to K. Martel.10.28.03.

Sobsey, M.D., A.P. Dufour, C.P. Gerba, M.W. LeChevallier, and P. Payment. 1993. Using a Conceptual Framework for Assessing Risks to Health From Microbes in Drinking Water. *JAWWA*, 85(3), 44-48.

South East Water Limited. 2002. *HACCP Plan*. Internal Document.

South East Water Limited. 1999. *Developing HACCP Systems Training for South East Water Limited*. Prepared by Australian Water Technologies and Quality Assurance Services.

Standards Australia/Standards New Zealand. 2004. *AS/NZS 4360:2004 Risk Management Standard, Third Edition*. Sydney, NSW, Australia: Standards Australia International Ltd.

Swerdlow, D.L., B.A. Woodruff, R.C. Brady, P.M. Griffin, S. Tippen, H.D. Donnell, Jr., E. Geldreich, B.J. Payne, A. Meyer, Jr., and J.G. Wells. 1992. A waterborne outbreak in Missouri of *Escherichia coli* O157:H7 associated with bloody diarrhea and death. *Ann Intern Med*, 117(10):812-9.

Swiss Gas and Water Industry Association. 2003. *W1002 Regulation-Recommendations for a Simple Water Quality Assurance System for Water Supplies*. Zurich, Switzerland: Swiss Gas and Water Industry Association.

U.S. Environmental Protection Agency. 1998. *National Primary Drinking Water Regulations: Interim Enhanced Surface Water Treatment Rule.* Federal Register, 63(241): 69478-69521.

U.S. Environmental Protection Agency. 1991. *Quality Assurance Glossary and Acronyms*. ORD-QA Management Staff. Washington, D.C.

World Health Organization (WHO). 2004. *Guidelines for Drinking-Water Quality, 3rd Edition.* Geneva, Switzerland: World Health Organization.

World Health Organization (WHO). 1997. *HACCP – Introducing the Hazard Analysis and Critical Control Point System*. Geneva, Switzerland: WHO.

World Health Organization (WHO). 1996. *Guidelines for Drinking-Water Quality, Second Edition*. Geneva, Switzerland: World Health Organization.

Yarra Valley Water, Ltd. 2005. HACCP Plan Version 8.1. Internal Document. Victoria, Australia: Yarra Valley Water Ltd.

ABBREVIATIONS

aka	also known as
AS	Standards Australia
AS/NZS	Australian/New Zealand Standard
AUD	Australian dollars
AWWA	American Water Works Association
AwwaRF	Awwa Research Foundation
C	Celsius
CCP	Critical control point
cfu	Colony forming units
CITEC	Australian Capital Territory Electricity and Water Corporation
Cl_2	chlorine
CRCWQT	Cooperative Research Center for Water Quality and Treatment
EMS	Environmental management system
EPA	Environmental Protection Agency
F	Fahrenheit
FAO	Food and Agricultural Organization
FDA	Food and Drug Administration
FSIS	Food Safety and Inspection Service
ft	feet
FTE	full-time equivalent
gal	gallons
GHP	Good hygiene practice
GIS	Geographic information system
GMP	Good manufacturing practice
gpm	gallons per minute
HACCP	Hazard Analysis and Critical Control Point
HPC	Heterotrophic plate count bacteria
HU	Hazen units
IBWA	International Bottled Water Association
IMS	Integrated management system
in	inches
ISO	International Organization for Standardization
km	kilometers
kPa	kilopascal
L	liters
LIMS	Laboratory Information Management System

m	meters
MCL	Maximum Contaminant Level
mg	milligram
MG	million gallon
min	minute
mL	milliliters
ML	megaliters
mm	millimeters
N/A	not applicable
NaOCl	sodium hypochlorite
NASA	National Aeronautics and Space Administration
NATA	National Association of Testing Authorities, Australia
NCCP	Non-critical control point
No.	Number
NSF	National Sanitation Foundation
NSW	New South Wales (State of New South Wales, Australia)
NTU	Nephelometric Turbidity Units
O & M	Operations and Maintenance
PAC	Project Advisory Committee
PHRMP	Public Health Risk Management Plan
PRV	pressure reducing valve
psi	pounds per square inch
QCP	Quality control point
SCADA	Supervisory Control and Data Acquisition
SOPs	Standard Operating Procedures
SSOPs	Sanitation Standard Operating Procedures
UCL	upper confidence level
US	United States
USA	United States of America
$\mu S\ cm^{-1}$	microSiemens per centimeter
USEPA	United States Environmental Protection Agency
UV	ultraviolet
VIC	Victoria (State of Victoria, Australia)
vs	versus
WF	Weighting factor
WHO	World Health Organization

IWA Publishing's authorised EU representative for General Product Safety Regulations is Diane D'Arras, 15 rue Duret, 75116 Paris, France, e-mail: safety@iwap.co.uk.

Printed and bound by CPI Group (UK) Ltd, Croydon, CR0 4YY
26/09/2026
02223459-0002